MATHEMATICAL METHODS OF ENVIRONMENTAL RISK MODELING

MATHEMATICAL METHODS OF ENVIRONMENTAL RISK MODELING

by

Douglas J. Crawford-Brown
University of North Carolina, at Chapel Hill

KLUWER ACADEMIC PUBLISHERS
Boston / Dordrecht / London

Distributors for North, Central and South America:
Kluwer Academic Publishers
101 Philip Drive
Assinippi Park
Norwell, Massachusetts 02061 USA
Telephone (781) 871-6600
Fax (781) 681-9045
E-Mail <kluwer@wkap.com>

Distributors for all other countries:
Kluwer Academic Publishers Group
Distribution Centre
Post Office Box 322
3300 AH Dordrecht, THE NETHERLANDS
Telephone 31 78 6392 392
Fax 31 78 6546 474
E-Mail <services@wkap.nl>

Electronic Services <http://www.wkap.nl>

Library of Congress Cataloging-in-Publication Data

Crawford-Brown, Douglas J.
Mathematical methods of environmental risk modeling / by Douglas J. Crawford-Brown.
p. cm.
Includes bibliographical references and index.

1. Environmental risk assessment—Mathematical models. I. Title

GE145 .C72 2001
363.7'02—dc21
2001035424

ISBN 978-1-4419-4900-4

Printed on acid-free paper.

Printed in the United States of America

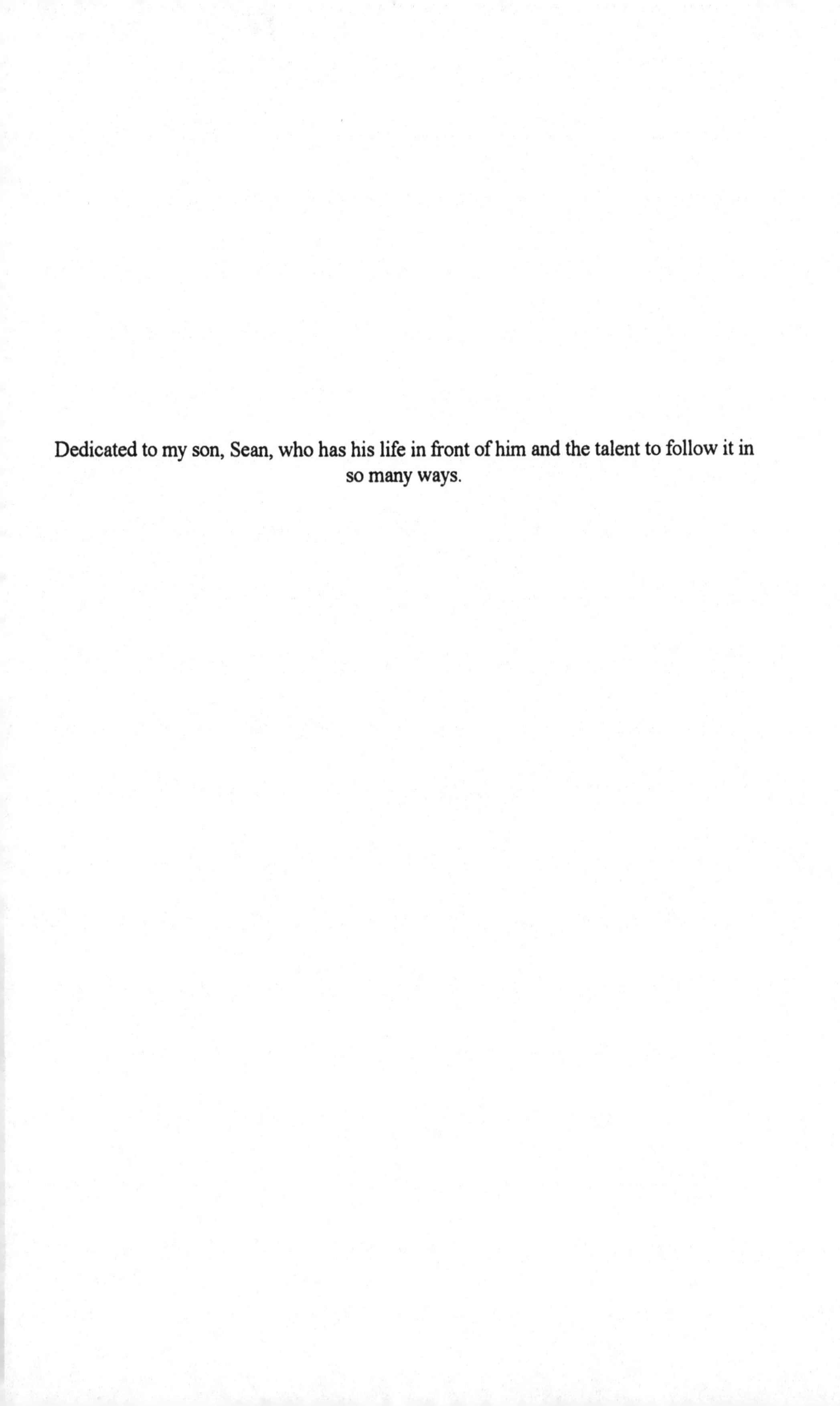

Dedicated to my son, Sean, who has his life in front of him and the talent to follow it in so many ways.

CONTENTS

Preface ix

Chapter 1. Fields, Spaces and States 1

1.1. The Concept of Fields and States 1
1.2. Scalar, Vector and Tensor Fields 5
1.3. The Gradient, Divergence and Curl of a Field 6
1.4. Translation, Rotation and Superposition of Fields 15

Chapter 2. Probability and Statistics 27

2.1. Decisions Under Variability and Uncertainty 27
2.2. Probability, Frequency, Confidence and Likelihood 29
2.3. Long-Term Frequentist and Bayesian Conceptions 32
2.4. Histograms and Probability Density Functions 35
2.5. Special Probability Density Functions 41
2.6. Correlation 46
2.7. Parameter Estimation and Measures of Model Quality 49
2.8. Error Propagation through Models 56

Chapter 3. Systems of Differential Equations 61

3.1. A Systems View of the Environment 61
3.2. Mass/Energy Balance and Conservation Laws 66
3.3. Linear Differential Equations 69
3.4. Systems of Differential Equations 71
3.5. Applications of Bernoulli's Method 75

Chapter 4. Laplace Transforms and Coupled Differential Equations 95

4.1. Coupled Systems and Feedback 95
4.2. Transforms 97
4.3. The Laplace Transform 99
4.4. The Inverse Laplace Transform 101
4.5. Applications of Laplace Transforms 102
4.6. Some Additional Laplace Transforms 122

Chapter 5. Matrix Methods and Spectral Analysis 125

5.1. Spectra in Environmental Problems 125
5.2. Back-elimination 126
5.3. Matrices 128
5.4. Augmented Matrices and Gauss-Jordan Elimination 134
5.5. Determinants Co-Factors, Minors and Inverses 136
5.6. Applications 140

Chapter 6. Numerical Methods and Exposure-Response 147

6.1. Exposure-Response Relationships 147
6.2. Numerical Integration 152
6.3. Numerical Solutions to Differential Equations: Euler's Method 157
6.4. Numerical Solutions to Differential Equations: Runge-Kutta Methods 163
6.5. The STELLA Modeling Software 167

Chapter 7. Monte Carlo Methods 175

7.1. Decisions Under Variability and Uncertainty 175
7.2. Analytic Methods 176
7.3. Monte Carlo Methods 179
7.4. Incorporating Model Uncertainty 187
7.5. Variability Between Geographic Regions and Subpopulations 194
7.6. Nested Variability and Uncertainty Analysis 196

Index 203

PREFACE

Isaac Newton placed mathematics firmly at the heart of science, and it remains there to this day. This means mathematics is not simply a tool to discover the world, but is also a language to express that discovery, to codify it, and to extrapolate experience to new realms. In the hands of someone as skilled as Newton, mathematics finds itself on an equal footing with empirical knowledge. Theories formulated in mathematics become the highest expression of understanding, revealing not only how the world behaves but the most fundamental laws governing that behavior. Use of mathematical models for prediction becomes more than the default tactic when data are unavailable; use of such models becomes the hallmark of rational prediction. The models move beyond mere means for prediction and become ends in and of themselves.

Given the importance of mathematics in the modern history of science (which I take to begin with Newton), it is not surprising that most science students gain a large dose of it during college, and continue to apply the methods in their careers. This is certainly true of the physical sciences and engineering, and is increasingly true of the biological sciences. Still, mathematics has played a more limited role in the education of risk analysts, particularly those analysts moving into the field from biological disciplines such as ecology and toxicology. A lack of training in mathematics remains a significant hurdle to the application of mathematical models in risk analysis, despite the potential of such models to guide empirical research and improve decisions. This has been less of a problem in exposure assessment, which is dominated by analysts from the physical sciences and engineering, but it remains a significant problem in exposure-response assessment, which is dominated by analysts trained in less mathematically inclined fields such as toxicology, medicine and ecology.

This book was written to fill the gap in mathematical expertise by anyone who wants to work in the area of environmental risk analysis, or to understand the basis for such analyses. It covers the techniques used to create, test and apply mathematical models of exposure assessment and exposure-response. As such, it is a book of applied mathematics rather than pure mathematics, and stems from the assumption that the reader will benefit most from seeing techniques "in action". It is the third in a series of books I have written on the topic of risk analysis, all published by Kluwer Academic Publishers. *Theoretical and*

Mathematical Foundations of Human Health Risk Analysis focused on the basic science of risk assessment, particularly as applied to public health. *Risk-Based Environmental Decisions: Methods and Culture* focused on applications of risk analysis to decisions. The present book focuses on mathematical techniques that arise most often in risk assessment, providing a wide range of examples using models employed typically in risk assessment to estimate human health risk.

Recognizing the diverse backgrounds of likely readers, I begin with only some minor requirements. The reader will need to review the basics of algebra and calculus, including the ideas of derivatives and integrals. Other than these most rudimentary tools, no formal training in advanced mathematics is presumed. All of the tools needed for at least the applications discussed in the examples may be learned purely from the material in the book. This has required that I focus more on the mechanics of the methods, and on "learning by doing", rather than on the fundamental theory behind those methods (e.g., there is little development of formal proofs for the methods, an issue that is central to purely mathematical texts). Illustrative examples have been chosen to cover the full range of problems that arise in environmental risk assessment, from pollutant source characterization, to transport and fate of pollutants in environmental systems, to pharmacokinetic and pharmacodynamic modeling, and on to health impacts.

Chapter 1 presents four of the most basic ideas in the mathematics of risk assessment: spaces, domains, fields and states. A central tenet of this book is that risk arises from the 4-dimensional (3 dimensions of space and one of time) distribution of pollutants in the body, which in turn depends on the 4-dimensional distribution of pollutants in environmental media, which in turn depends on the 4-dimensional distribution of sources of that pollution. The idea of fields is illustrated through consideration of gaussian dispersion models for atmospheric pollutants and the resulting exposures of geographically inhomogeneous populations. In addition, methods for translation, superposition and addition of fields are covered.

Chapter 2 presents the concepts of probability and statistics needed to estimate parameters for models and to build stochastic models. These same techniques may be applied (and are applied here) to issues of variability and uncertainty analysis. Also included in this chapter is a discussion of the distinction between deterministic and stochastic models, and brief discussion of simple parameter estimation methods. The example used throughout the chapter is uncertainty and variability analysis for the exposures identified in Chapter 1, and the development of statistical exposure-response relationships.

Chapter 3 explores the solution of systems of ordinary differential equations using one particular technique (Bernoulli's solution) that has proven effective in models typically encountered in exposure assessment and pharmacokinetics. These equations arise often in systems analysis and, if the system does not become too complicated, they may be solved analytically. The primary examples used are transfer of a pollutant through environmental systems and transformation of a pollutant through chemical reactions.

Chapter 4 focuses on LaPlace Transforms and their application to coupled sets of differential equations. Such equations represent systems in which there is flow of a pollutant or a biological state both forward and backward. LaPlace Transforms allow development of an analytic solution to this problem, at least for relatively simple systems. Pharmacokinetics and the transformation of cells between states of health, with repair of damage, are used as the examples.

Chapter 5 contains a discussion of matrix methods useful in simplifying complex environmental models, and for unfolding parameter values from complex sets of data. These techniques are particularly useful in input-output analysis, where the input may be determined from the sensitivity of a system and the response of that system. The primary example used is unfolding the state of the environment from information on the response of multiple species and their underlying sensitivity matrices. In addition, matrix methods are related to the issue of surface water flow in a watershed.

Chapter 6 moves to numerical techniques when the analytic methods discussed in previous chapters may no longer be applied (a situation that arises routinely in environmental science due to the complexity of environmental systems). The chapter considers numerical methods for integration and for the solution of differential equations, using the examples of multi-media modeling.

Chapter 7 focuses on Monte Carlo modeling for analysis of variability and uncertainty. It ends with consideration of nested variability and uncertainty analyses, and the way such information is used in selecting risk-based policies.

CHAPTER 1
Fields, Spaces and States

1.1. The Concept of Fields and States

Models allow us to both understand and predict phenomena. In the case of models used in environmental health risk assessment, the phenomena are those of the relationship between sources of risk (e.g. a factory), the state of the environment (e.g. the state of the atmosphere), and various states of health. The predictions tell us what will happen to both the state of the environment and the state of health under well-described circumstances such as an experiment or introduction of an environmental policy. If models also embody understanding, they tell us *why* these events happen in the order in which they do, based on an understanding of the principles governing the phenomenon. Environmental models are most successful when they embody both prediction and understanding, and so this book focuses on both aspects of a model.

At the heart of environmental modeling lies the idea of a *state*. States are both physical and mathematical. They describe the conditions existing in some region of space (typically a three-dimensional region defining a volume, but we will see examples of two-dimensional regions defining an area) and at some point in time. These conditions, in turn, can be described by three characteristics:

- *Property*- the physical (chemical, biological, etc) aspect of a region of space that is being described. The property might, for example, be the concentration of a pollutant, or the temperature of a gas, or the intensity of sunlight.

- *Quantity*- the numerical value associated with this property. The concentration of the pollutant at a particular point in the space might be 3. The range of values over which the quantity may vary is the *domain* for that quantity; the domain for most environmentally relevant quantities is 0 to infinity, shown as $[0,\infty]$.

- *Unit*- the metric through which the quantity is being expressed. In the example above, the property is the concentration of a pollutant, the quantity is 3, and the unit might be grams per liter (g/L).

A *field* is both a physical entity existing in a space, and a mathematical description of that entity. Each field is characterized by a specific property, quantity and unit. The property then is referred to as the *field property*, and the quantity as the *field quantity*. There is a field quantity associated with each point in the space, although the field should not be confused with the space containing it. An example of a field is shown in Figure 1.1, which is a surface representing the density of population over a geographic region. The space in this example is two-dimensional, given by the plane formed by the x- and y-axes. The third (z) axis shows the quantity characterizing the field property; the units of this particular field property are people per square kilometer. Note that this quantity varies throughout the space; not all points in the field have the same quantity (although they have the same property and units). The field, therefore, is an *inhomogeneous field.* A *homogeneous field* is one in which the quantity associated with the property being measured does not vary for the different points in the space. An example of a homogeneous field is shown in Figure 1.2.

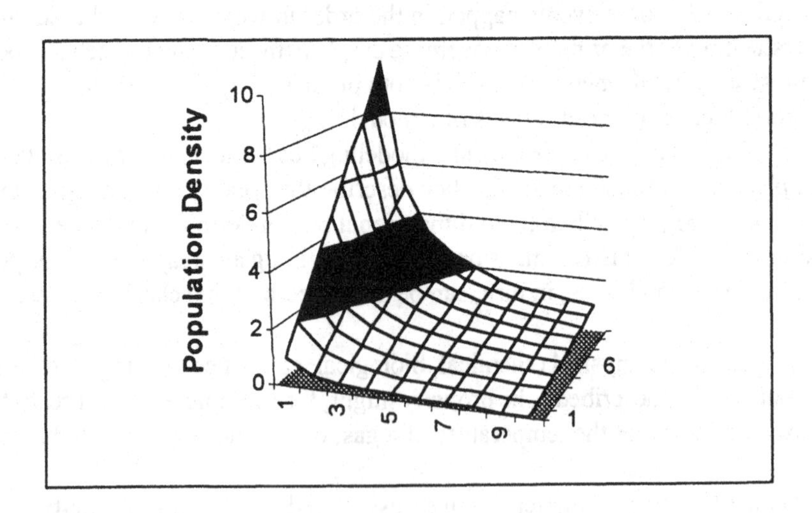

Figure 1.1. An example of a field; in this case the field property is density of a population throughout a geographic region (units of people/km^2). The space on which the field is defined is 2-dimensional, and the field itself is inhomogeneous.

A field will be shown in this book as F(x,y,z,t), indicating that the field may be a function of the 3 spatial coordinates (x,y,z) and of time (t). Since the field may change in time, we will refer to the *evolution of the field.* Figure 1.3 shows an example of a field at two points in time, indicating that the field has evolved, since the quantity characteristic of at least one spatial point has changed between these two points in time.

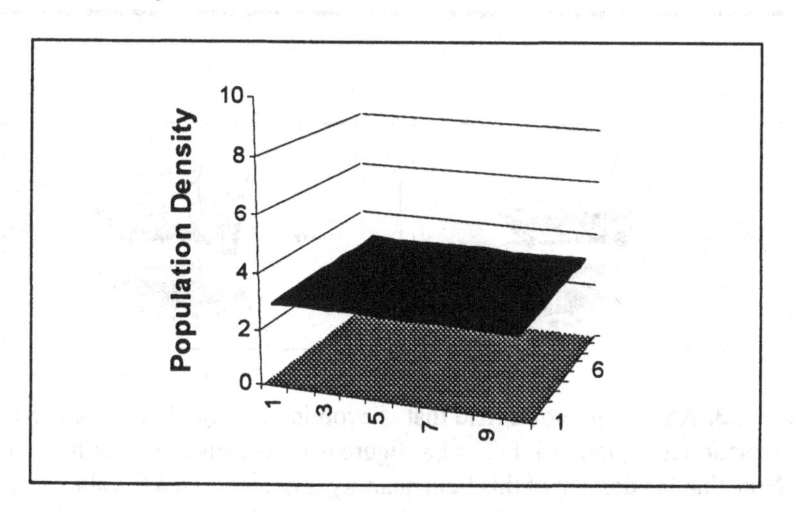

Figure 1.2. An example of a homogeneous field with value 3 at all points in the space. The field property and field units are the same as in Figure 1.1.

Fields are both mathematical and physical. There really is a concentration of environmental tobacco smoke in the air of a room containing a cigarette smoker, and the resulting concentration at each point in space constitutes the physical field of atmospheric concentration. At the same time, there is a mathematical description of this field that allows the risk assessor to determine quantitatively the property at each point in the field. Underlying this mathematical description is a *field equation* by which one may calculate the field quantity at any point in space and moment in time. For Figure 1.3, the field equation is:

(1.1) $$F(x,y,t) = \sin(x^t)/y$$

Example 1.1. Consider the field in Equation 1.1 at the point (x,y) = (10,20) and the time t = 5. The field quantity at this point in space and time is:

$$F(10,20,5) = \sin(10^5)/20 = 0.0018$$

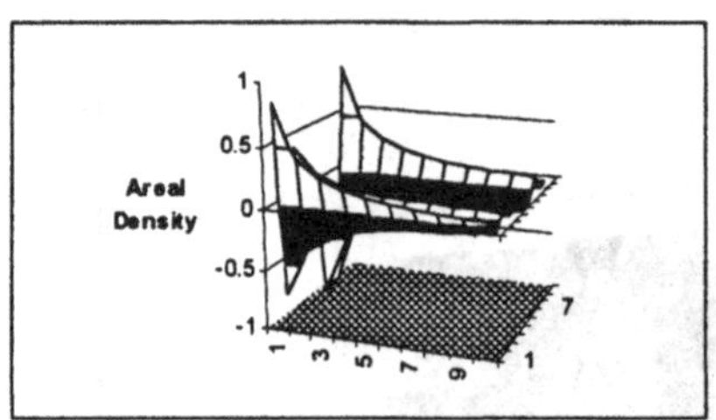

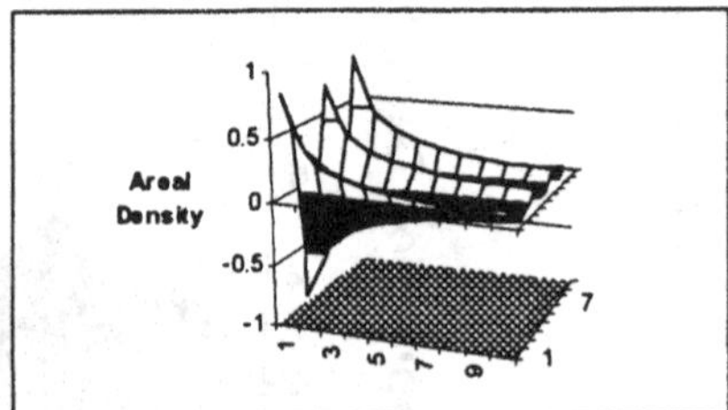

Figure 1.3. An example of a field that is evolving in time. In this example, the field equation is Equation 1.1. The left figure is for t=2 and the right figure is for t=5. Note that the domain of this field quantity is not restricted to values of 0 and above. This might, for example, be a field of the difference between the rate of cancer at a point in a geographic region and the average for the region, with a negative value meaning that point has a rate below the average.

The *state of the environment* is the collection of all fields relevant to the particular risk assessment being performed [1]. As a simple example, imagine performing a risk assessment for environmental tobacco smoke (ETS). Within a room, there might be a field of atmospheric concentration ($\mu g/cm^3$) for the ETS. Since smoke consists in part of particles, there will be settling of the smoke onto surfaces such as the top of a table. This will produce another field, the two-dimensional field of particle density ($\mu g/cm^2$). Since a person might be exposed to both of these fields, the state of the environment will include both fields. In this case, the state of the environment at some moment in time is described by displaying the fields of atmospheric concentration and particle density. Since the fields themselves evolve in time, so will the state of the environment. Note also that the state of the environment will be inhomogeneous if at least one of the fields also is inhomogeneous.

To carry the example further, and to complete the scope of the models considered in this book, consider that the field of atmospheric concentration depends on the sources of the pollutant (in this case, cigarettes being smoked) and the movement and fate of the smoke after being released from the source. This source may also be described by a field that varies in space and time. We will shown it here as the three-dimensional field S(x,y,z,t). It produces the three-dimensional field of atmospheric concentration, C(x,y,z,t), which in turn produces the two-dimensional field of particle density D(x,y,t). Both C(x,y,z,t) and D(x,y,t) then are responsible for changing the state of health, which we might show as the field of density of health effects (e.g. number of asthmatic attacks per km^2) throughout a geographic region, H(x,y,t); the H stands for "health" in this example. The models used in environmental health risk assessment share the characteristics of this example, with one field producing another. In this example, S(x,y,z,t) produces C(x,y,z,t), which produces D(x,y,t), which in turn along with C(x,y,z,t) produces H(x,y,t). Everywhere one turns in environmental health risk assessment one finds fields and the field equations through which they are both explained and predicted.

1.2. Scalar, Vector and Tensor Fields

Fields come in three varieties within environmental health risk assessment, depending on the nature of the field property [2]. *Scalar fields* describe properties that have only a magnitude. Examples are concentration, pressure, temperature and density. The field quantity is a single number at each point in space and time. All of the figures shown so far (Figures 1.1 to 1.3) are examples of scalar fields.

Vector fields describe properties that have a quantity with components in each of the three spatial directions. In general, these components are shown as $F_x(x,y,z,t)$, $F_y(x,y,z,t)$ and $F_z(x,y,z,t)$. An example is the field of velocity, V(x,y,z,t), in which each spatial point is characterized by a component in the x direction, $V_x(x,y,z,t)$; a component in the y direction, $V_y(x,y,z,t)$; and a component in the z direction, $V_z(x,y,z,t)$. The *magnitude* of a vector quantity, shown as $|F(x,y,z,t)|$, is calculated through the formula:

$$(1.2) \qquad |F(x,y,z,t)| = [F_x^2 + F_y^2 + F_z^2]^{0.5}$$

Separate field equations may apply to each of the three components.

Tensor fields describe properties that have more than three components. Tensor fields are not typically encountered in environmental risk assessment, with the possible exception of models of earthquakes. The classic tensor field in the physical sciences is one describing stresses on planes [3]. The stresses at a particular point in space can be described by a 9 component field specifying the shearing stress on the x-y plane in the x direction; the shearing stress on the x-y plane in the y direction; the shearing stress on the y-z plane in the z direction; the shearing stress on the y-z plane in the y direction; the shearing stress on the x-z plane in the x direction; the shearing stress on the x-z plane in the z direction; the normal stress on the x-y plane in the z direction; the normal stress on the y-z plane in the x direction; and the normal stress on the x-z plane in the y direction. The result is a 3 x 3 matrix of stresses, or a tensor field. Tensor fields are not considered to any large extent in this book, other than in the context of the curl in Section 1.3.

Example 1.2. Consider a vector field in which $V_x(1,3,5,2)$ is 2 m/s; $V_y(1,3,5,2)$ is 4 m/s; and $V_z(1,3,5,2)$ is 0 m/s. The magnitude of the velocity (i.e. the speed) at this point in space and moment in time, using Equation 1.2, is:

$$|V(x,y,z,t)| = [2^2 + 4^2 + 0^2]^{0.5} = 4.47 \text{ m/s}$$

Movement is restricted in this example to the x-y plane since there is no velocity in the z direction. The angle of movement, Φ, with respect to the x-axis is found by noting that it must satisfy the following relationship:

$$\tan(\Phi) = V_y/V_x = 4/2 = 2$$

$$\text{or} \quad \Phi = \arctan(2) = 1.107 \text{ radians or } 63.5 \text{ degrees}$$

Note: π (or 3.14) radians is 180 degrees

1.3. Gradient, Divergence and Curl of a Field

There is a wide variety of secondary properties of fields that, rather than describing the field property at each point in space, describe how the field changes in space or in time. The first such property is a *gradient*, which describes how the magnitude of the field quantity changes (increases or decreases) as one moves along any given axis direction [4]. To illustrate, consider the following simple field equation (the field does not evolve in time, so there is no reference to t):

$$F(x,y,z) = 2x + 3y + z^2 \tag{1.3}$$

A gradient at a point (x,y,z) may be calculated as the slope of the field in any of the three directions at that point. This gives rise to three components of the gradient for the field in Equation 1.3:

(1.4a) $$dF(x,y,z)/dx = 2$$

(1.4b) $$dF(x,y,z)/dy = 3$$

(1.4c) $$dF(x,y,z)/dz = 2z$$

Note: Throughout this book, the product of two quantities is shown either by placing them together (e.g. 2z in Equation 1.4c) or separated by the symbol "•" if simple placement together is confusing. The reason for choosing "•" rather than "x" to indicate multiplication is to avoid confusion with the coordinate x.

Notice that in this example, the gradients in the x and y directions do not depend on the (x,y,z) coordinates, while the gradient in the z direction does not depend on the (x,y) coordinates but does depend on the z coordinate. If the gradient is negative in a given direction, the field quantity decreases as one moves out the positive direction of the axis; if the gradient is positive, the field quantity increases as one moves out the positive direction of the axis. Note that all three gradients in Equations 1.4a to 1.4c are positive.

Gradients play an important role in environmental models because the movement of pollutants often is related to the gradient. If a pollutant moves by diffusion, it will move from a region of high concentration to a region of low concentration. If a pollutant moves under pressure, it will move from a region of high pressure to a region of low pressure. This movement is "down the gradient", with the term "down" meaning in the direction in which the gradient is negative. In particular, the direction and magnitude (speed) of flow can be determined from the same vector operations introduced in Section 1.2. Specifically, the magnitude of the gradient (which is proportional to the magnitude of the flow) is found from Equation 1.2:

(1.5) $$\mathrm{grad}[F(x,y,z)] = [(dF(x,y,z)/dx)^2 + (dF(x,y,z)/dy)^2 + (dF(x,y,z)/dz)^2]^{0.5}$$

grad[F(x,y,z)] in Equation 1.5 is usually what is meant by "the" gradient at some point in space (x,y,z).

Example 1.3. Consider the field defined by Equation 1.3, and the three components of the gradient in Equations 1.4a to 1.4c. "The" gradient at the point (x,y,z) = (10,8,5) is:

$$grad[F(10,8,5)] = [(dF(10,8,5)/dx)^2 + (dF(10,8,5)/dy)^2 + (dF(10,8,5)/dz)^2]^{0.5}$$

$$= [2^2 + 3^2 + (2 \bullet 5)^2]^{0.5} = 10.6$$

The flow of a pollutant under a gradient can be illustrated by considering a pressure field in the atmosphere, and the resulting flow of air. Applying Darcy's Law to atmospheric flow [5], one finds:

(1.6) $$V_i(x,y,z) = -k_i \bullet dP(x,y,z)/di$$

where $V_i(x,y,z)$ is the velocity of the flow in the ith direction, k_i is a constant that increases as the material is easier to move (it is essentially the inverse of the resistance to flow), and $dP(x,y,z)/di$ is the gradient of the pressure field in the ith direction at the point (x,y,z). The negative sign on the right hand sign of the equation exists because a negative gradient along the positive direction of an axis results in a flow in the positive direction along that axis (remembering that the air will flow "down the gradient", or from high pressure to low pressure). With a negative value for dP_i/di, the right hand side will become positive when the pressure gradient is negative along the positive axis (since k_i is always a positive number) and the velocity will be positive. Note that if the pressure gradient is positive in a given direction i, the velocity will be negative, since pressure is increasing as one moves outwards along the positive axis in that direction, and the flow is in the opposite direction (i.e. the negative direction along that axis).

There generally will be three components to flow, corresponding to the three spatial dimensions. Equation 1.6 then has three components:

(1.7a) $$V_x(x,y,z) = -k_x \bullet dP(x,y,z)/dx$$

(1.7b) $$V_y(x,y,z) = -k_y \bullet dP(x,y,z)/dy$$

(1.7c) $$V_z(x,y,z) = -k_z \bullet dP(x,y,z)/dz$$

The magnitude of the velocity (or the speed of flow) can be calculated from

Equation 1.3:

(1.8) $$|V(x,y,z)| = [V^2_x(x,y,z) + V^2_y(x,y,z) + V^2_z(x,y,z)]^{0.5}$$

If $k_x = k_y = k_z = k$, and if Equations 1.7a to 1.7c are inserted into Equation 1.8, the result is:

(1.9) $$|V(x,y,z)| = -k \bullet [(dP(x,y,z)/dx)^2 + (dP(x,y,z)/dy)^2 + (dP(x,y,z)/dz)^2]^{0.5}$$

Finally, using Equation 1.5, Equation 1.9 can be re-written as:

(1.10) $$|V(x,y,z)| = -k \bullet \text{grad}[(P(x,y,z)]$$

In other words, the field of velocity is related to the gradient of another field, the field of pressure.

The flow of air under a pressure field illustrates another useful concept, that of *isopleths*. Consider a simple pressure field that depends on the values of x and y, but not on the value of z. For simplicity, let that pressure field be defined by the field equation $P(x,y,z) = x + y$. This pressure field is shown in Figure 1.4 Rather than plotting in three dimensions, the field is shown as the two dimensional projection down onto the x-y plane (since the pressure is not a function of z, the same projection applies for all values of z). In this figure, all points (x,y) characterized by the same numerical value of the field property have been connected by a line. The resulting lines represent isopleths, or lines of equal pressure. If a point (x,y,z) is located in this plane, the flow will be in the direction that carries the air to the next lower isopleth in the shortest distance. Note that this also characterizes the direction of the steepest negative gradient, since this direction moves from one isopleth to the next in the smallest distance. Hence, the change in pressure per unit distance is the largest in this direction. In general, at any point (x,y,z), this direction will correspond to a line which is normal (i.e. at 90 degrees) to the tangent of the isopleth at that point.

The second property of a field we will consider is *divergence* [4], shown here as divF(x,y,z). The divergence is a property of vector fields (gradients can apply to scalar fields) and, in the case of the flow of an environmental medium is a measure of how a material either "spreads out" or "compresses" as it flows through the environment. If the divergence is positive, the flow causes the material to spread out into a larger volume, or "diverge" as it flows; negative divergence represents the case where the material compresses as it flows. By definition, an incompressible fluid must have a divergence of 0. The divergence can be inhomogeneous over the space of a field, even changing sign within that space.

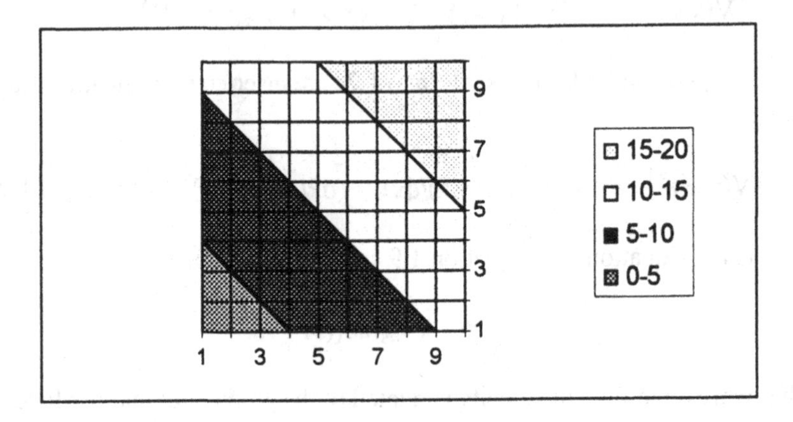

Figure 1.4. An example of isopleths, representing lines of equal pressure (pressures in torr are indicated in the box to the right of the figure). In this example, the air will flow from the upper right to the lower left of the figure, always in a direction normal to the isopleths. For example, consider the point (x,y) = (7,8). Air will flow in the direction of the point (x,y) = (6,7) and from there to the point (x,y) = (5,6) and so on.

This book continues with the example of flow of material in an environmental medium (air, water, etc). The general formula for divergence is:

$$(1.11)\quad divV(x,y,z) = dV_x(x,y,z)/dx + dV_y(x,y,z)/dy + dV_z(x,y,z)/dz$$

Substituting Equations 1.7a to 1.7c into Equation 1.11, one obtains:

$$(1.12)\ divV(x,y,z) = -k_x \bullet dP(x,y,z)/dx - k_y \bullet dP(x,y,z)/dy - k_z \bullet dP(x,y,z)/dz$$

Note that divergence is a scalar quantity since it has a single numerical value at each point in the space. To understand the physical meaning of the divergence, consider a very simple pressure field in which the pressure changes only along the x-axis and the gradient of the pressure is constant. Let P(x,y,z) = 2x. For this pressure field, using Equation 1.7a, the velocity in the x-direction is:

$$V_x(x,y,z) = -k_x \bullet dP(x,y,z)/dx = -2k_x$$

Note that the velocity is constant at all points (x,y,z), and that it is pointed backwards along the x-axis (i.e. it is a negative velocity).

Using Equation 1.11, and noting that the component of divergence in the y and z-direction is 0, the divergence for this velocity field is:

$$\mathrm{divV}(x,y,z) \;=\; dV_x(x,y,z)/dx + 0 + 0 \;=\; d(-2k_x)/dx \;=\; 0$$

For this velocity field, the velocity is the same at all points and so the fluid simply moves parallel the x-axis with no compression or expansion. The divergence in such a case is 0.

Example 1.4. Consider a pressure field defined by $P(x,y,z) = x^2 + y^2 + z^2$ (this is not a physically realistic field, but is chosen to simplify the derivations). Let k be the same numerical value for all three directions and equal to 3 (units of m/s per torr/m). Using Equations 1.7a to 1.7c, the three components of velocity for the flow of the air at the point (x,y,z) = (5,7,2) are:

$$V_x(5,7,2) \;=\; -k \bullet dP(5,7,2)/dx \;=\; -3 \bullet 2x \;=\; -3 \bullet 2 \bullet 5 \;=\; -30 \text{ m/s}$$
$$V_y(5,7,2) \;=\; -k \bullet dP(5,7,2)/dy \;=\; -3 \bullet 2y \;=\; -3 \bullet 2 \bullet 7 \;=\; -42 \text{ m/s}$$
$$V_z(5,7,2) \;=\; -k \bullet dP(5,7,2)/dz \;=\; -3 \bullet 2z \;=\; -3 \bullet 2 \bullet 2 \;=\; -12 \text{ m/s}$$

And using Equation 1.5:

$$|V(5,7,2)| \;=\; \{V^2_x(5,7,2) + V^2_y(5,7,2) + V^2_z(5,7,2)]^{0.5} \;=$$

$$[(-30)^2 + (-42)^2 + (-12)^2]^{0.5} \;=\; 52.99 \text{ m/s}$$

Note that, given this pressure field (which increases symmetrically as one moves out in any direction), flow is inwards towards the origin (0,0,0). This can be seen in part by noting that the flow at (x,y,z) = (5,7,2) is negative for all three components, or back towards the origin.

Now consider a more complicated case in which the velocity is a function of x. For example, let $V_x(x,y,z) = 5x$, which means the velocity is positive parallel to the x-axis and increasing as one moves out the x-axis. Intuitively, it should be clear that the "leading edge" of a packet of the fluid (i.e. the edge furthest out the x-axis) is moving faster than the "trailing edge". The result is that the packet should be stretching as it moves. Calculating the divergence, again using Equation 1.11, yields:

$$\mathrm{div}V(x,y,z) = dV_x(x,y,z)/dx + 0 + 0 = d(5x)/dx = 5$$

Note that this value is positive, meaning the material in the packet is diverging or spreading out into a larger volume as it moves. If one considers the case where $V_x(x,y,z) = -5x$, or the velocity is negative, the divergence is:

$$\mathrm{div}V(x,y,z) = dV_x(x,y,z)/dx + 0 + 0 = d(-5x)/dx = -5$$

which implies the packet of material is diverging as it moves in the negative x direction, which is the same as converging with respect to the positive x direction.

If the velocity field also has non-zero derivatives with respect to the y and z directions, it is possible for a positive divergence along one axis to be cancelled by a negative divergence along another. In other words, it is not necessary that the contribution to the divergence be 0 along all three axes to produce a case in which the overall divergence is zero. What this would mean physically is that the shape of a packet of material might change as it flows but, if the material is incompressible, the overall volume containing the packet does not change.

Example 1.5. Consider the same velocity field as in Example 1.4, where:

$$V_x(x,y,z) = -k \bullet dP(5,7,2)/dx = -3 \bullet 2x$$
$$V_y(x,y,z) = -k \bullet dP(5,7,2)/dy = -3 \bullet 2y$$
$$V_z(x,y,z) = -k \bullet dP(5,7,2)/dz = -3 \bullet 2z$$

The three terms contributing to the divergence for this field are:

$$dV_x(x,y,z)/dx = -6$$
$$dV_y(x,y,z)/dy = -6$$
$$dV_z(x,y,z)/dz = -6$$

So the divergence is:

$$\mathrm{div}V(x,y,z) = -6 - 6 - 6 = -18$$

Note that this is negative divergence, so the material is compressing as it moves.

The final concept considered in this section is that of the *curl* of a velocity field [4]. Curl is a measure of the tendency of a velocity field to produce "swirling" motions. To understand the curl of such a field, it is useful to think of a rigid bar spinning around an axis, with one end of the bar fixed at some point (x,y). Imagine the bar rotating in the counter-clockwise direction around the point. The point is located on the x-y plane (such as on this sheet of paper). For such a bar, the "outer edge" of the bar, or point on the bar furthest from the axis of rotation, is moving at a higher linear velocity than the innermost point (the linear velocity is higher for the outermost point, even though the angular velocity must be the same for all points along the bar since the bar is rigid). We will employ the right-hand rule: using the right hand, point all the fingers except the thumb in the direction of rotation of the bar. The direction in which the thumb points is the direction of the curl. If this paper is the x-y plane, and if the positive z-axis is defined as pointing towards the reader, and if the bar is swirling in the counter-clockwise direction, it should be evident that the curl in the z direction is positive since the thumb is pointing out from the paper and towards the reader (i.e. along the positive direction of the z-axis).

Consider now the velocity of such a bar as one moves along the x-axis. This velocity has two components: a velocity in the x direction, $V_x(x,y,z)$, and a velocity in the y direction, $V_y(x,y,z)$. We will define the positive y direction as being towards the top of the page, and the positive x direction as being towards the right hand edge of the page. Let the bar rotate around until it is pointed along the x-axis in the positive direction (i.e. towards the right hand edge of the page). What happens to $V_y(x,y,z)$ as one moves along the positive direction on the x-axis? The magnitude of $V_y(x,y,z)$ increases, and the sign is positive (remember, the bar is rotating counter-clockwise), so the gradient of $V_y(x,y,z)$ in the x direction, or $dV_y(x,y,z)/dx$, must be positive. A positive curl in the z direction corresponds to a positive gradient of $V_y(x,y,z)$ in the x direction, or a positive value of $dV_y(x,y,z)/dx$. The higher the positive gradient of $V_y(x,y,z)$ in the x direction, the safter the bar is rotating.

What of the velocity in the x direction, and its contribution to the curl in the z direction? Let the same bar rotate until it is pointed directly up the y-axis (i.,e. towards the top of the page). Since the bar is rotating counter-clockwise, the velocity in the x direction must be negative. What happens to $V_x(x,y,z)$ as one moves along the positive direction on the y-axis? The magnitude of $V_x(x,y,z)$ increases but the sign is negative, so the gradient of $V_x(x,y,z)$ in the y direction, or $dV_x(x,y,z)/dy$, must be negative. A positive curl in the z direction corresponds to a negative gradient of $V_x(x,y,z)$ in the y direction, or a negative value of $dV_x(x,y,z)/dy$. Since both $dV_y(x,y,z)/dx$ and $dV_x(x,y,z)/dy$ can contribute to the curl in the z direction, the curl in the z direction is defined as:

(1.13a) $\quad \text{curl}_z V(x,y,z) = dV_y(x,y,z)/dx - dV_x(x,y,z)/dy$

Using identical arguments for the curl around the y-axis and the curl around the z-axis, one obtains:

(1.13b) $\quad \text{curl}_x V(x,y,z) = dV_z(x,y,z)/dy - dV_y(x,y,z)/dz$

(1.13c) $\quad \text{curl}_y V(x,y,z) = dV_x(x,y,z)/dz - dV_z(x,y,z)/dx$

The curl is a vector quantity and so it has three components that can be added vectorially such as in Equation 1.2. The higher the curl at a point, the greater the tendency of the material to swirl around that point rather than to move in a straight line.

Caution: Equations 1.13a to 1.13c were developed with respect the particular directions in which the positive x-, y-, and z-axes were defined. The signs on the two terms on the right hand sides of Equations 1.3a to 1.3c may change if the directions of the positive axes are defined differently.

Example 1.6. Consider the velocity field from Examples 1.4 and 1.5. Calculating the 6 derivatives needed to define the curl around the point (x,y,z) as shown in Equations 1.13a to 1.13c:

$dV_y(x,y,z)/dx = 0$ (since V_y does not depend on x)
$dV_x(x,y,z)/dy = 0$ (since V_x does not depend on y)
$dV_z(x,y,z)/dy = 0$ (since V_z does not depend on y)
$dV_y(x,y,z)/dz = 0$ (since V_y does not depend on z)
$dV_x(x,y,z)/dz = 0$ (since V_x does not depend on z)
$dV_z(x,y,z)/dx = 0$ (since V_z does not depend on x)

and the three components of the curl are:

$$\text{curl}_z V(x,y,z) = 0 - 0 = 0$$
$$\text{curl}_x V(x,y,z) = 0 - 0 = 0$$
$$\text{curl}_y V(x,y,z) = 0 - 0 = 0$$

Note that the curl is 0 at all points, so there is no swirling motion.

1.4. Translation, Rotation and Superposition of Fields

Environmental risk assessments routinely require calculations of fields under many different scenarios. For example, the assessor might need to calculate the concentration of a pollutant in the atmosphere in the space surrounding an industrial facility under many different combinations of control strategies (different air filters, etc). It might also be necessary to calculate the field of concentration in the air, water and soil, and then to combine these in some way to estimate the total risk to the population. Rather than performing these calculations many times, one for each scenario, there often are ways to perform a single set of calculations and then combine them in different ways to simulate the different scenarios. This section reviews those techniques. The example used throughout is of a pollutant released from an industrial facility into the air, and the transport of that pollutant through the air.

Before developing the mathematics of the example, it is necessary to consider the mechanisms by which pollutants move through the environment. They are discussed here with respect to dispersion in the atmosphere, but the same mechanisms apply to movement in any environmental medium (air, water, soil, etc) and to movement through the human body. If a volume of gas leaves the smokestack of a facility, it will move through the atmosphere primarily by the following four processes [6]:

- *Buoyancy*- the movement of a substance upwards due to differences in density. Materials of lower density (such as a hot gas leaving a smokestack) will rise above the denser material (the cool air surrounding the stack). The rising gas will expand and cool, and will continue to rise until its temperature and density are the same as that of the surrounding air. The direction of movement is upwards in the atmosphere. Buoyancy generally will be greater as the *lapse rate*, or rate of change of air temperature with altitude, is larger and negative (i.e. buoyancy will be greater if the temperature of the surrounding air tends to decrease rapidly with increases in altitude). The height a parcel of pollutant (such as a hot gas) rises into the atmosphere after being released is the *effective stack height*, H. It is a function of the temperature of the gas, the velocity at which it is released from the stack, the height of the stack, the diameter of the stack, and the lapse rate of the atmosphere. A higher effective stack height generally results in greater dispersion of a pollutant into the atmosphere, resulting in the pollutant being carried large distances.

- *Carriage*- the movement of a substance in streams of the environmental medium. For air dispersion, carriage is on the wind currents, and the direction of movement of the pollutant is in the direction of the currents carrying it. This direction is not constant, and so carriage in the atmosphere is described by a *wind rose* showing the fraction of time the wind blows in a particular direction (actually, by convention, wind roses often show the fraction of time the wind blows *from* a particular direction). An example wind rose might provide the following information (which will be used in subsequent examples in this section): East (0.2,3); Northeast (0.1,1); North (0.1,1); Northwest (0.2,2); West (0,1); Southwest (0.1,2); South (0.05,1); Southeast (0.25,3). The first number in each set is the fraction of time the wind blows *towards* that direction, and the second is the wind speed when the wind is blowing in that direction. For the North, for example, the wind blows in that direction 10% of the time, and at an average speed of 1 m/s.

- *Diffusion*- the random movement of atoms and molecules of a pollutant due to the fact that the temperature of the pollutant is above zero and, hence, the kinetic energy is not zero. The movement is uniform in all directions, so a parcel of gas will tend to spread as a sphere of increasing diameter as time increases. The rate of this diffusion depends on the size and shape of the atoms and molecules, their temperature, and on the viscosity of the medium through which they are diffusing. Generally, the diffusion will increase as particle size decreases, as the shape becomes more spherical, as temperature increases, and as viscosity decreases.

- *Sedimentation*- the movement of a pollutant towards a center of gravity due to the weight of the atoms or molecules. The direction for atmospheric pollutants is downwards towards the surface of the earth. In general, heavier particles will settle most quickly since they are more difficult to keep aloft by the air currents, unless their increased surface area causes additional friction that prevents settling.

Consider the fate of a hot pollutant leaving a smokestack. It will rise by buoyancy until it cools to the temperature of the surrounding air; it will move horizontally with the wind currents; it will diffuse in all directions; and it may sediment to the ground. We will consider only a gas here, so sedimentation can be ignored.

Figure 1.5 shows a bird's-eye view of a geographic region (a two dimensional space) containing such a smokestack at the origin (x,y) = (0,0). Imagine the wind blowing entirely from west to east at a speed of u. As the pollutant moves downwind, it forms a plume and spreads out due to diffusion. This plume can be described by a centerline that is at the middle of the plume and follows the x-axis. Walking outwards from the source along this centerline, and stopping at some distance x from the source (x is called the *downwind distance*), imagine turning left or right and walking along a line 90 degrees from the centerline. This is a *crosswind walk*. The distance from the point on the centerline where the turn occurred, and any point at which a concentration is to be estimated during the crosswind walk, is the *crosswind distance*, y.

As a simple approximation for this discussion, the concentration one would measure in performing this crosswind walk would decrease as y increases. If one walks first in one direction from the centerline, and then returns and walks in the opposite direction, the resulting concentrations encountered at ground-level can be described approximately by a *gaussian function* (the normal, or bell-shaped, curve so familiar from statistics). As with all such curves, it is necessary to specify the *standard deviation*, σ, characterizing the gaussian distribution. This standard deviation generally will increase as one considers points further downwind, since there is more time for diffusion to occur before the plume reaches these points.

The concentration of the pollutant at the ground level at any point (x,y) downwind then can be described by the *Gaussian Plume Model* [7]:

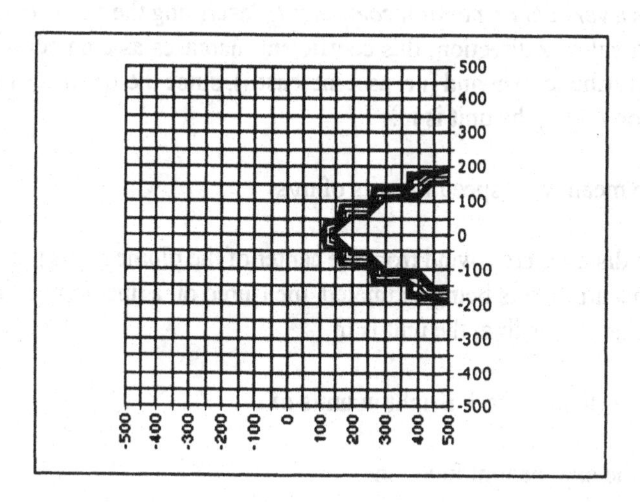

Figure 1.5. A bird's-eye view of a geographic region with a smokestack at the origin (0,0). The dark lines show the concentration isopleths of a plume containing the pollutant released from the stack. The "non-smooth" nature of these lines is due to the plotting software, and not to the calculation of concentration. Note that the width of the plume is increasing as the plume moves further from the source. The wind is blowing towards the east.

$$(1.14)\ C(x,y) = \{Q / (2\pi \bullet \sigma_y(x) \bullet \sigma_z(x) \bullet u)\} \bullet \exp\{-0.5(y^2/\sigma^2_y(x)) - 0.5(H^2/\sigma^2_z(x))\}$$

where:

- $C(x,y)$ is the ground-level concentration of the pollutant at a point (x,y) assuming the source is at $(0,0)$ and the wind blows only along the x-axis; the units are g/m^3.
- Q is the release rate of the pollutant from the stack, in units of g/s.
- $\sigma_y(x)$ is a *horizontal dispersion coefficient*, describing the extent of diffusion in the cross-wind direction; this coefficient increases as a parcel of pollutant moves further downwind (i.e. as x increases), since there is more time for the diffusion to act; the unit is m.
- $\sigma_z(x)$ is a *vertical dispersion coefficient*, describing the extent of diffusion in the vertical or z direction; this coefficient increases as a parcel of pollutant moves further downwind (i.e. as x increases), since there is more time for the diffusion to act; the unit is m.
- u is the mean wind speed in units of m/s.
- y is the distance crosswind from the center of the plume to the point at which the concentration is being estimated, measured on a line perpendicular from the plume center-line; the unit is m.
- H is the effective stack height in units of m.
- exp is the exponential function.

An example of the field of concentration predicted by Equation 1.14 is shown in Figure 1.6. The gaussian shape can be seen by looking at the "lip" at the far right, or eastern-most, edge of the figure. This figure was produced under the same conditions as those assumed in Figure 1.5, with Figure 1.5 being the projection of Figure 1.6 down onto the x-y plane (i.e. onto the ground). Note that the highest concentration is not directly next to the source, since buoyancy causes an effective stack height that can be well above ground-level. The pollutant must travel some distance downwind before diffusion and settling brings it back to the ground where people can be exposed.

The field shown in Figure 1.6 results from the source when the wind is blowing due east. Over a sufficiently long period of time (e.g. several months), the wind will change direction many times, moving through the entire wind rose. When the wind is in directions other than to the east, the resulting concentration field for the pollutant in the air will be different than that shown in Figure 1.6. Let $C_i(x,y)$ be the concentration at point (x,y) when the wind blows in the ith direction.

In some cases, the fields produced when the wind blows in the different directions can be added. In other words, the average concentration at point (x,y) is the weighted sum of the concentration fields produced during the different wind directions, with the weighting being the fraction of time the wind blows in each direction. The principles of *superposition* and *additivity* state that the different fields may be superimposed and added in this way. If f_i is the fraction of time the wind blows in the ith direction, the time-averaged concentration field will be given by:

$$(1.15) \qquad C(x,y) = \Sigma f_i \bullet C_i(x,y)$$

where Σ indicates the summation is over all wind directions. In other words, the concentration at a given point (x,y) is determined for each of the fields, the different fields are superimposed, and the time-weighted average obtained at that point (x,y) using Equation 1.15. Substituting Equation 1.14 into 1.15 yields:

$$(1.16) \qquad C(x,y) = \Sigma\{Q \bullet f_i / (2\pi \bullet \sigma_y(x) \bullet \sigma_z(x) \bullet u_i)\} \bullet \exp\{-0.5(y^2/\sigma^2_y(x)) - 0.5(H^2/\sigma^2_z(x))\}$$

Equations 1.14 and 1.16 are valid so long as the source is located at (0,0). What is to be done if the source is not located there, but rather at (x_s,y_s)? The first step in solving this problem is to imagine that the same concentration field surrounding the source will be produced as in Equations 1.14 and 1.16, but with the field shifted along the x-axis a distance x_s and along the y-axis a distance y_s. We could calculate the concentration field under the assumption that the source is located at (0,0) as in these two equations, and then ask which point in the concentration field produced by the new source location corresponds to a given point (x,y) in the field produced by a source located at (0,0). This is a problem of the *translation* of fields. The field produced with the source at (0,0) is calculated once and stored, and then translated when a new source location is considered.

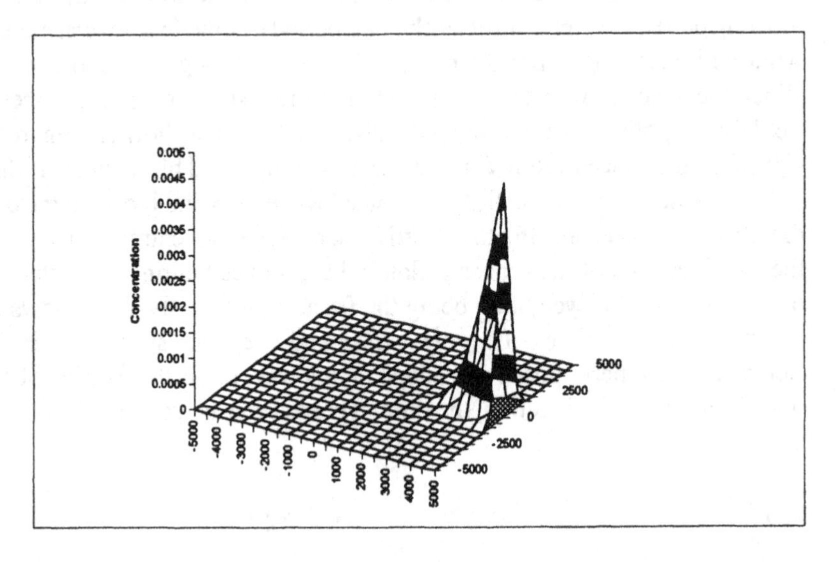

Figure 1.6. The air concentration field for a pollutant released into the atmosphere from a source located at (x,y) = (0,0). The wind is blowing only towards the east, or the right-hand edge of the plane. The concentration is in units of g/cm^3. The approximate gaussian shape may be seen at the right-most edge of the field.

Example 1.7. Consider a source located at (x,y) = (0,0). The wind blows constantly towards the east at a velocity of 1 m/s. The pollutant is released at a rate of 1000 g/s and at an effective stack height of 50 m. The goal is to estimate the concentration of the pollutant at ground-level for the point (x,y) = (100m,30m). The value of x in Equation 1.16 is 100 m, and the value of y is 30 m. For this value of x, σ_y is 20 m and σ_z is 10 m. From Equation 1.16:

$$C(100,30) = \Sigma\{Q \bullet f_i / (2\pi \bullet \sigma_y(x) \bullet \sigma_z(x) \bullet u_i)\}$$

$$\bullet \exp\{-0.5(y^2/\sigma^2_y(x)) - 0.5(H^2/\sigma^2_z(x))\}$$

$$= \{1000 \bullet 1 / (2\pi \bullet 20 \bullet 10 \bullet 1)\} \bullet \exp\{-0.5(30^2/20^2) - 0.5(50^2/10^2)\}$$

$$= 9.6 \times 10^{-7} \text{ g/m}^3$$

Consider a point (x,y) in the field produced when the source is at (x_s,y_s). In Equations 1.14 and 1.16, the value of x in the equation is not really the x-coordinate of the point at which the concentration is being estimated. It is, instead, the downwind distance associated with that point. In other words, it is the distance in the x direction from the source to the point where the concentration is being calculated. This distance is $(x-x_s)$ when the wind blows due east. When x_s is 0 (as was assumed in generating Equations 1.14 and 1.16), this downwind distance simply equals x, which is why x_s did not appear in those equations. Similarly, the value of y in the equation is not really the y-coordinate of the point at which the concentration is being estimated. It is, instead, the crosswind distance associated with that point (i.e. the distance along the perpendicular line from the centerline of the plume to the point at which the concentration is being estimated). This distance is $(y-y_s)$ when the wind blows due east. When y_s is 0 (as was assumed in generating Equations 1.14 and 1.16), this crosswind distance simply equals y, which is why y_s did not appear in those equations.

Both x and y in Equations 1.14 and 1.16 should be replaced by $(x-x_s)$ and $(y-y_s)$, respectively, to account for the translation of the field that must take place if the source is not at (0,0). Using Equation 1.16, which is the most complete form of the model (including all directions of wind flow), yields:

$$(1.17)\ C(x,y) = \Sigma\{Q \bullet f_i / (2\pi \bullet \sigma_y(x-x_s) \bullet \sigma_z(x-x_s) \bullet u_i)\} \bullet \exp\{-0.5((y-y_s)^2/\sigma^2_y(x-x_s)) - 0.5(H^2/\sigma^2_z(x-x_s)\}$$

Note that σ_y and σ_z must now be evaluated at the value of $(x-x_s)$ rather than simply at x. This process of translation of fields also is called *mapping*. The procedure is as follows:

1. Calculate the concentration field assuming the source is at (0,0). Call the points in that space (x^*,y^*).

2. Create a new space (a new x-y plane) with the source at (x_s,y_s). Call the points in that space (x,y).

3. Pick a point (x,y) in the second space. Ask which point (x^*,y^*) in the first space contains the concentration that would have been produced in the second space at the point (x,y) with the source at (x_s,y_s). This will be the point in the first space corresponding to $(x^*,y^*) = (x-x_s,y-y_s)$.

4. Map, or translate, the concentration at this point $(x-x_s,y-y_s)$ in the first space onto the point (x,y) in the second space.

If this is done for all points (x,y) in the second space , with one corresponding point (x*,y*) from the first space yielding the concentration associated with each point (x,y) in the second space, the field produced in the first space will have been mapped onto the second space. Note that this process means the model can be run once with the source at (0,0) and the resulting field translated under all new combinations of source locations.

Return now to the issue of the wind rose and the fact that the wind blows in multiple directions over a period of time. This was shown in Equations1.15 and 1.16. As with translation, it is necessary to calculate the field produced under one set of conditions (with the wind blowing due east) and then map this field onto the space of interest (where the wind points in many directions). This is a process of *field rotation*.

The problem of rotation for the gaussian model is shown in Figure 1.7. The wind is blowing at an angle Φ with respect to the x-axis (or east direction). The task is to find the concentration at the point (x,y) in the plane when the wind blows in that direction. Equation 1.14 is valid, with the caveat that x in that equation refers to the downwind distance when moving in the plume center and y refers to the crosswind distance when moving from the plume center outwards (on the perpendicular line) to the point at which the concentration is being estimated. If Φ is not zero, x and y in Equation 1.14 are not the same as the (x,y) coordinates of the point at which concentration is being estimated. We need a relationship between the downwind distance (which we will call x_Φ) and the values of the coordinates (x,y) and the angle of the wind, Φ. We also need a relationship between the crosswind distance (which we will call y_Φ) and the values of the coordinates (x,y) and the angle of the wind, Φ. These relationships will allow the rotation of the field, or the mapping. The derivations of these relationships are provided below.

(1.18) $$x_\Phi \cos(\Phi) + y_\Phi \sin(\Phi) = x$$

and

(1.19) $$x_\Phi \sin(\Phi) - y_\Phi \cos(\Phi) = y$$

We can rearrange Equation 1.18 to give:

(1.20) $$x_\Phi \cos(\Phi) = x - y_\Phi \sin(\Phi)$$

or

(1.21) $$x_\Phi = (x - y_\Phi \sin(\Phi))/\cos(\Phi)$$

Substituting Equation 1.21 into Equation 1.19, and noting that $\sin^2(\Phi) + \cos^2(\Phi) = 1$, one obtains:

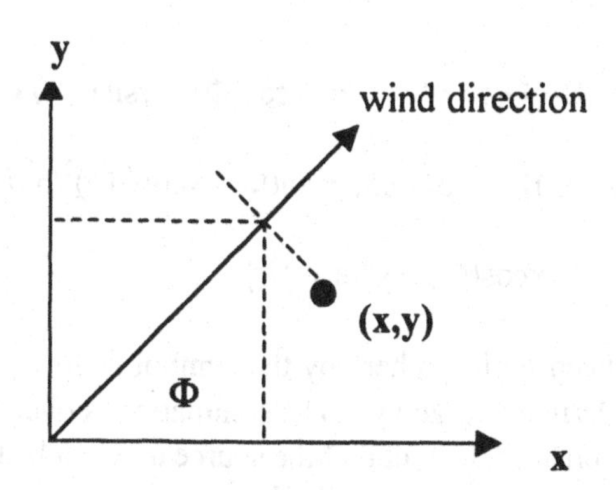

Figure 1.7. The geometry of rotation for the concentration field. The wind blows at an angle Φ with respect to the x-axis. The diagonal dashed line is perpendicular to the centerline of the plume. The other two dashed lines (one vertical and one horizontal) are perpendicular to the x- and y-axis, respectively. The angle between the vertical dashed line and the diagonal dashed line also is Φ. The concentration is being estimated at the point (x,y). The crosswind distance y_Φ is the distance from the centerline to the point (x,y) along the diagonal dashed line. The downwind distance x_Φ is the distance from the origin to the intersection of the three dashed lines.

(1.22) $$y_\Phi = x\sin(\Phi) - y\cos(\Phi)$$

Now substitute the solution for y_Φ into Equation 1.18 to get:

(1.23) $$x_\Phi \cos(\Phi) + (x\sin(\Phi) - y\cos(\Phi))\sin(\Phi) = x$$

Again after a few steps involving the fact that $1 - \sin^2(\Phi) = \cos^2(\Phi)$, one finds:

(1.24) $$x_\Phi = x\cos(\Phi) + y\sin(\Phi)$$

Equations 1.22 and 1.24 provide the relationships needed to rotate the field through any angle Φ. Note that if Φ is 0 degrees, x_Φ equals x, as expected. Note also that if Φ is 0 degrees, y_Φ equals -y. This seems to be incorrect, but then note that in Equation 1.14 y appears only as y^2, and so it is only the magnitude of y_Φ that is needed, and not the sign.

Inserting Equations 1.22 and 1.24 into Equation 1.16, the following general relationship for the concentration of the pollutant at ground level at a point (x,y) is found:

$$(1.25)\quad C(x,y) = \Sigma\{Q \bullet f_\Phi / (2\pi \bullet \sigma_y(x\cos(\Phi) + y\sin(\Phi)) \bullet \sigma_z(x\cos(\Phi) + y\sin(\Phi)) \bullet u_\Phi)\} \bullet \exp\{-0.5((x\sin(\Phi) - y\cos(\Phi))^2/\sigma^2_y(x\cos(\Phi) + y\sin(\Phi))) - 0.5(H^2/\sigma^2_z(x\cos(\Phi) + y\sin(\Phi)))\}$$

Note that f_i has been replaced here by the symbol f_Φ to be consistent with the symbols used in deriving x_Φ and y_Φ. The summation is over all values of Φ.

Finally, consider the fact that the source may not be located at the origin of the space. It may be necssary to both rotate and translate the field. Under translation, x becomes $(x-x_s)$ and y becomes $(y-y_s)$. Inserting these two relationships into Equation 1.25, the most general form of the model is obtained:

$$(1.26)\quad C(x,y) = \Sigma\{Q \bullet f_\Phi / (2\pi \bullet \sigma_y((x-x_s)\cos(\Phi) + (y-y_s)\sin(\Phi)) \bullet \sigma_z((x-x_s)\cos(\Phi) + (y-y_s)\sin(\Phi)) \bullet u_\Phi)\} \bullet \exp\{-0.5(((x-x_s)\sin(\Phi) - (y-y_s)\cos(\Phi))^2/\sigma^2_y((x-x_s)\cos(\Phi) + (y-y_s)\sin(\Phi))) - 0.5(H^2/\sigma^2_z((x-x_s)\cos(\Phi) + (y-y_s)\sin(\Phi)))\}$$

Again, the summation is over all values of Φ, or all wind directions contained in the wind rose.

Equation 1.26 illustrates four principles from this section: translation of fields, rotation of fields, superposition of fields (one field for each wind direction), and additivity of fields. This latter principle of additivity is not always valid. If there are non-linearities in the phenomena being modeled, such as a case in which a chemical reacts with itself while in the air, additivity cannot be assumed. Models for dealing with this case of non-additivity are beyond the scope of this book.

Consider also the fact that Equation 1.26 consists of superimposing and adding a finite number of wind directions (usually 8, as in Example 1.7). The calculated concentrations will be highest along the centerline of the plume, dropping off on either side of this centerline. The result will be a pattern of predicted concentrations showing ridges and valleys, as can be seen in Figure 1.8. These ridges and valleys are the result of replacing a continuous phenomenon (the wind actually swings around continuously) by a discrete approximation (only 8 wind directions). These artifacts can be removed in part by increasing the number of wind directions (from 8 to 16 to 32, etc), but this requires more information and larger computation time. If the ridges and valleys are not removed, it is important that the model not be used to estimate concentrations at isolated points in the space, but rather be used to obtain average concentrations over regions of the space that are larger than the artifacts. For example, one might average the predicted concentrations in Figure 1.8 over several grid blocks.

Example 1.8. Consider the wind rose given previously as East (0.2,3); Northeast (0.1,1); North (0.1,1); Northwest (0.2,2); West (0,1); Southwest (0.1,2); South (0.05,1); Southeast (0.25,3). Consider the same source as in Example 1.6, but with the source located at $(x_s,y_s) = (100,100)$. The pollutant is released at a rate of 1000 g/s and at an effective stack height of 50 m. Using Equation 1.26, the following field is obtained:

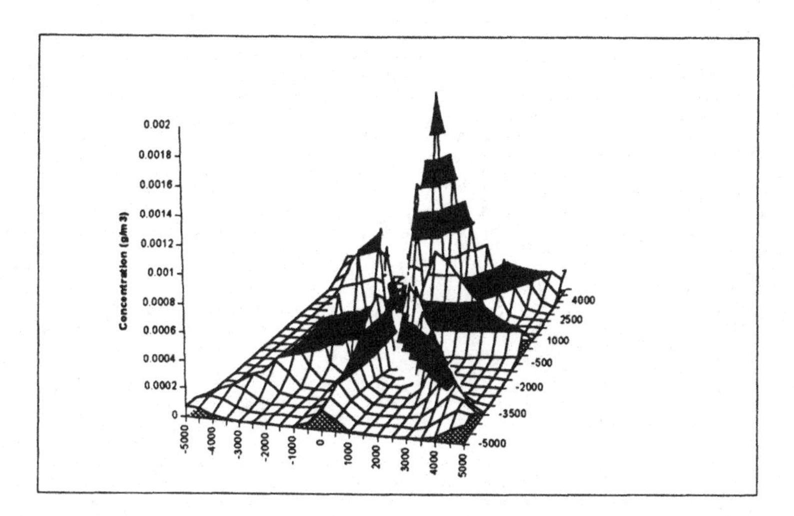

Figure 1.8. The atmospheric concentration field for a pollutant released into air with the wind rose shown above. Note the ridges and valleys discussed previously, which are in part an artifact of the use of 8 wind directions.

References

1. D. Crawford-Brown, *Theoretical and Mathematical Foundations of Human Health Risk Analysis*, Kluwer Academic Publishers, Boston, 1997.
2. P. Kundu, *Fluid Mechanics*, Academic Press, San Diego, 1990.
3. L. Scipio, *Principles of Continua with Applications*, John Wiley and Sons, Inc., New York, 1967.
4. H. Schey, *Div, Grad, Curl and All That*, W.W. Norton and Company, New York, 1997.
5. G. Hornberger, J. Raffensperger, P. Wiberg and K. Eshleman, *Elements of Physical Hydrology*, The Johns Hopkins University Press, Baltimore, 1998.
6. D. Crawford-Brown, *Risk-Based Environmental Decisions: Methods and Culture*, Kluwer Academic Publishers, Boston, 2000.
7. J. Henry and G. Heinke, *Environmental Science and Engineering*, Prentice-Hall, Inc., Upper Saddle River, NJ, 1996.

CHAPTER 2
Probability and Statistics

2.1. Decisions Under Variability and Uncertainty

Environmental phenomena, and the methods by which these phenomena are measured, are characterized by variability and uncertainty. In developing environmental models, several sources of variability and uncertainty appear routinely and complicate both predictions and the decisions on which they are based:

- *Variability of physical properties in space and/or time*. Section 1.1 introduced the concept of an inhomogeneous field. In such fields, the field quantity varies across the space at some point in time. These same fields may evolve in time, and so there is variation in the field quantity for a particular point in space at different times. A full description of the environmental field requires characterizing this spatial inhomogeneity and temporal evolution.

- *Variability of physical properties across a population.* Two individuals exposed to the same state of the environment may have different states of health. This might, for example, be because of intersubject variability in breathing rates, body mass, etc. A full description of the risk from an environmental field requires characterization of this variation between individuals.

- *Random sampling by a measurement method.* Measurement methods provide completely reliably measures of field quantities only when performed over a very long period of time with a large number of samples. In practice, a finite, manageable, sample size must be taken. This leads to variation in measurements of a field quantity, even at a single point in space. This variation is not a property of the field itself (that component of variation is accounted for by the first bullet above), but rather of the measurement system.

- *Stochasticity in environmental processes*. Some phenomena (such as those controlled by quantum processes) are inherently random in nature, or *stochastic*. It is not possible to predict such phenomena with complete precision, even when presented with perfect knowledge, since the outcome is random rather than deterministic. This component of stochasticity differs from the first bullet in that this component results from a stochastic process, while the first results from a deterministic process with parameter values that vary in space and time. It also differs from the second bullet in that this component refers to stochasticity in the phenomenon itself, while the second bullet results from randomness in the measurement process performed on that phenomenon.

- *Uncertainty in predictions due to the existence of competing models*. In practice, the risk assessor often is faced with a number of models, all purporting to describe the phenomenon in question. Each model will have some degree of support in the available evidence, and so each model is a possible rational basis for predictions. A full risk assessment depicts the differences in predictions from the available models, and assigns some rational measure of confidence to each such prediction.

- *Uncertainty in predictions due to a lack of confidence in the available science*. It is possible that the modeler is not completely confident that *any* of the models predicts and/or explains the phenomenon accurately, or even that the suite of existing models bounds the truth. It is necessary to ask, therefore, whether the entire field of study might be insufficiently developed to warrant making any reliable predictions and/or explanations.

These sources of variability and uncertainty move risk-based environmental decisions away from a simple prediction of the effects of a policy and the comparison of those effects against goals. It is necessary, instead, to consider the variation in these effects across an exposed population, and the uncertainty in predicting this variation.

As a result of variability and uncertainty, decisions based on environmental risk assessment usually are rooted in the idea of a *margin of safety*. This margin reflects a concern for the *precautionary principle* which states that, in the absence of perfect knowledge, risk estimates should be selected that are likely, if incorrect, to produce overestimates of the risk to individuals in a population. The most general form of a decision goal or task given to the risk assessor and consistent with the precautionary principle can be stated as follows:

Locate a policy (e.g. state of the pollution source) that will protect at least X% of the exposed population against unreasonable risks, where one is at least Y% confidence that X has not been overestimated.

This formulation of a risk-based goal places variability and uncertainty squarely at the center of modeling performed in support of risk-based decisions. Specifically, the issue of estimating the fraction of the population protected (i.e. X%) is related to the concept of *intersubject variability* of risk (i.e. variation in risk across populations). The issue of the confidence in risk predictions (i.e. Y%) is related to the concept of *uncertainty*. For a model to be fully useful in a risk assessment, it will be necessary to understand both the variability in predictions across a population and the uncertainty in the model predictions.

This means models used in environmental risk assessment must be subjected to *statistical analysis*. Originally, the term "statistics" referred simply to the collection of data in areas such as population and economic studies [1]. Over time, it came to refer to the entire process of analyzing these data, summarizing them, using them to describe and communicate bodies of information (*descriptive statistics*), and using them to go beyond the data and draw tentative conclusions about a larger population (*inferential statistics*) [2]. The methods discussed in this chapter focus primarily on descriptive statistics, with an eye towards characterizing variability and uncertainty in the fields that make up the state of the environment and the state of health.

2.2. Probability, Frequency, Confidence and Likelihood

While the sources of variability and uncertainty described in Section 2.1 may be treated using similar mathematical methods, there are important conceptual differences between these sources. These differences are embodied in four concepts that appear regularly in statistical analyses of environmental models: probability, frequency, confidence and likelihood.

In this book, *probability* refers to a stochastic process in which the outcome of a measurement may be treated as a random variable. This might arise either because the phenomenon being measured is stochastic, or because the measurement process is stochastic. In any event, repeated measurements produce a distribution of outcomes, and knowledge of the phenomenon and/or measurement process, however complete, will not remove this inherent source of uncertainty. The reason is that a probability is an irreducible characteristic of each individual entity being measured. For example, under the rules of quantum mechanics, the location of a particle in space is stochastic and can only be described by a probability. It is not possible to improve the prediction of the location beyond some limit, since the probability is an inherent property of the entity being measured (the particle) and not of the measurement process alone. Throughout this book, a probability is shown as p(x), interpreted as the probability that the field quantity has a given numerical value, x. For example, p(7) might be the probability that a field quantity such as concentration has a value of approximately 7. We will provide a more rigorous definition of "approximately" in Section 2.4.

Frequency refers to a case in which the outcome varies across repeated measurements because of systematic differences in the phenomenon being measured. For example, there is variation in the weights of individuals in a population. These weights are not stochastic, since each individual has a specific weight that could, in principle, be determined accurately. The weights do, however, differ between individuals and this intersubject variability can be described by stating the fraction of individuals whose weights fall into some category (e.g. from 50 to 55 kg). The frequency of an outcome is the fraction of times this outcome occurs in repeated measurements. Throughout this book, a frequency is shown as f(x), interpreted as the fraction of individuals in a population for which the field quantity has a particular numerical value, x. For example, f(7) might be the fraction of individuals (people, points in a space, etc) characterized by a field quantity of approximately 7. Again, we will provide a more rigorous definition of "approximately" in Section 2.4.

Note that probability and frequency can be related at times. Consider an event characterized by a probability, such as the location of a particle in space. While the property being measured (location) truly is stochastic, and can be described for any one particle by a probability, it also is possible to consider a large number of identical particles and perform measurements of the locations of each of these. These measurements will result in a frequency of outcomes; e.g. 80% of the time the particle is located in a sphere within 10^{-12} m of a particular point in space. In this case, frequency is an expression of an underlying probability, and results from repeated measurements on a stochastic process. Given a large enough sample size, and an unbiased measurement method, the frequency will equal the probability, or $f(x) = p(x)$.

It also is possible to have a frequency without a probability (at least in the sense of probability adopted here). Returning again to the issue of weights of people in a population, there is no sense in which a given individual has a probability of weighing 50 kg. The person either does or does not weigh 50 kg. Frequency in this case refers to variation of the weight across different individuals. Probability applies to individuals in a population, while frequency is a property of the population.

Both probability and frequency are objective properties of either an individual (probability) or a population of individuals (frequency). At times, one is interested in expressing a property of the person making claims about some belief (which might be a belief about probabilities and/or frequencies, but is not restricted to such beliefs). The goal in this case is to express the level of *confidence* in the belief. The term confidence is kept separate from probability and frequency since it refers to a property of the subject (the person having the belief) while the latter are properties of the object of that belief. In philosophical terms, probability and frequency are ontological properties, while confidence is an epistemological property.

It is possible to express confidence in terms or units similar to those used in probability and frequency. A statement might be made that one is 30% confident that the weight of an individual is approximately 50 kg. This confidence might even be based on objective information such as probability and/or frequency. For example, one might know that the frequency of individuals with a weight of approximately 50 kg is 30%, and argue that the degree of confidence in the belief should correspond to the fraction of time it will be true in a random sample. Confidence need not, however, reflect only frequencies, and it is always a subjective property and not an objective property. We will return to this issue in more detail in Section 2.3.

Consider the case in which it is necessary to estimate the weight of an individual. A measurement is performed and indicates 50 kg. It is recognized, however, that the measurement process is in some sense stochastic. There is a chance, therefore, that the true weight is 55 kg, or 30 kg, etc. We might ask the following questions:

How likely is it that, IF the true weight were 30 kg, our measurement process would have produced 50 kg; how likely is it that, IF the true weight were 55 kg, our measurement process would have produced 50 kg; etc?

We could show these questions in general as $L(50 \mid x)$, which is translated as: the *likelihood* (L) that 50 would be the result of the measurement *given that* the real value is x. To make the expression even more general, we will write $L(y \mid x)$, or the likelihood that y would be the result of a measurement given that the real value is x. For the two examples in the italics above, the likelihoods are $L(50 \mid 30)$ and $L(50 \mid 55)$

Likelihoods are related to probabilities and/or frequencies. The likelihood that 50 kg is measured given that the real weight is 30 kg is precisely the probability that, or frequency with which, a measurement process applied to a 30 kg weight will produce an erroneous measurement of 50 kg. Likelihoods cannot be calculated without knowing the underlying probabilities and/or frequencies. Still, the terms probability, frequency and likelihood are kept distinct because of the specialized application of probability within the calculation of likelihoods.

Example 2.1. Consider the problem of determining the concentration of a pollutant in a lake. The measurement system is characterized by uncertainty. In particular, 20% of the time it produces a result that is too high by a factor 2, 30% of the time it produces a result that is too low by a factor of 2, and 50% of the time it produces a result exactly equal to the actual concentration. A measurement is performed and the concentration is indicated to be 10 g/m^3. Using the definition of likelihood, the following 3 likelihoods may be calculated:

The likelihood that the true concentration is 20 g/m^3 is the probability that the measurement result is too low by a factor of 2:

$$L(10|20) = L(y/2|y) = 0.3$$

The likelihood that the true concentration is 5 g/m^3 is the probability that the measurement result is too high by a factor of 2:

$$L(10|5) = L(2y|y) = 0.2$$

The likelihood that the true concentration is 10 g/m^3 is the probability that the measurement result is exactly equal to the real concentration:

$$L(10|10) = L(y|y) = 0.5$$

2.3. Long-Term Frequentist and Bayesian Conceptions

There is a particularly strong conceptual separation between probability and frequency on the one hand and confidence on the other. This is due to the difference between *objective* and *subjective* properties. The result has been the development of two schools of thought on the nature of probabilistic claims, which can be seen in the discussions between Einstein and Bohr on the interpretation of quantum mechanics. To Bohr, quantum mechanics, depicting quantities such as the position of a particle as stochastic and inherently random, is the ultimate description of reality. The probabilities to which it refers are an objective property of that reality; they are *ontological probabilities*. The probability of a particle being at a particular location equals the fraction of time it will be found at that location given an infinite (or at least very large) number of identical measurements. Bohr's conception of probability falls within the *long-term frequentist* school of thought [3], where probability is measured by the frequency of occurrence of some event in an infinitely large sample.

Einstein thought of quantum mechanics as a confession of uncertainty due to incomplete understanding. At least his early argument was that the probabilities expressed in quantum mechanics resulted from ignorance of *hidden variables*, or factors that, once understood, would lead to deterministic calculations. The probabilities would disappear as soon as these hidden variables were located. This led him to his statement that, in effect, God does not play dice with the universe.

Einstein's beliefs fall at least in part within the *Bayesian* school of thought [4], in which probabilities are expressed as measures of confidence since they are expressions of the strength of belief in claims, and not of any randomness in the underlying reality those beliefs are intended to describe. The Bayesian approach does not fully reject the idea of ontological probabilities resulting from purely stochastic phenomena (an idea Einstein *did* reject in his discussions with Bohr). It does, however, reject the idea that such probabilities are the *only* factors to take into account when forming a measure of rational confidence. To a Bayesian, one must also consider information at hand prior to making any measurements that are themselves characterized by probabilities.

Consider the case of estimating the value of a field quantity at some point in space, such as the concentration C(x,y,z). An example of such a process was shown in Example 2.1. A measurement is performed and a value of $C_m(x,y,z)$ is obtained (the subscript m stands for "measurement"). Under the long-term frequentist approach, the probability that the true concentration is $C_t(x,y,z)$, where the subscript t stands for "true", is calculated entirely from the likelihood as shown in Example 2.1; i.e. from $L(C_m(x,y,z) \mid C_t(x,y,z))$.

There may, however, have been some prior understanding of the situation. Suppose, for instance, that the source of the pollutant is known (it is a paper plant near the lake), and that this information suggests the concentration should be somewhere "in the vicinity of" $C_p(x,y,z)$, where the subscript p stands for "prior information" (it is prior because it is available before the measurement is performed). Rather than relying solely on the measurement $C_m(x,y,z)$ and on the associated likelihood function $L(C_m(x,y,z) \mid C_t(x,y,z))$, one might want to incorporate this prior information into any statements of confidence. In the simple Bayes approach (there are more complex approaches beyond the scope of this book), the confidence that the true value is $C_t(x,y,z)$ is found from [5]:

$$(2.1) \qquad c(C_t) = Pr(C_t) \bullet L(C_m \mid C_t) / \Sigma\, Pr(C_t) \bullet L(C_m \mid C_t)$$

where the summation in the denominator is over all possible true values of C_t. $Pr(C_t)$ is the prior for C_t, or the degree of confidence one has, prior to the measurement, that the true value is C_t. While the likelihood function in Equation 2.1 is an objective property of the phenomenon being measured and of the measurement system, the priors are subjective and so the confidence generated by the Bayesian approach also is subjective.

Note what happens if there is complete ignorance before the measurement is made. The priors $Pr(C_t)$ then are equal for all possible values of C_t, and the priors in the numerator and denominator cancel. This leaves: in Equation 2.1:

$$c(C_t) = L(C_m | C_t) / \Sigma L(C_m | C_t) \quad (2.1b)$$

In other words, confidence becomes identical to the calculation of likelihoods, and the Bayesian and long-term frequentist approaches converge.

Example 2.2. A paper plant emits a pollutant into a lake. There are only three possible concentrations in this simple example: 1 g/m^3, 5 g/m^3 and 10 g/m^3. Based on prior information, a risk assessor is 30% confident that the true concentration is 1 g/m^3, 20% confident that the true concentration is 5 g/m^3, and 50% confident that the true concentration is 10 g/m^3. A measurement is performed and indicates 15 g/m^3. Likelihoods for this measurement system are:

$L(15 | 1) = 0.1$
$L(15 | 5) = 0.3$
$L(15 | 10) = 0.6$

The confidence that the true concentration is 1, 5 or 10 g/m^3 is calculated using Equation 2.1a as:

$$c(1)= Pr(1) \bullet L(15 | 1)/\{Pr(1) \bullet L(15 | 1)+ Pr(5) \bullet L(15 | 5)+ Pr(10) \bullet L(15 | 10)\}$$

$$= 0.3 \bullet 0.1 / \{0.3 \bullet 0.1 + 0.2 \bullet 0.3 + 0.5 \bullet 0.6\} = 0.08$$

$$c(5)= Pr(5) \bullet L(15 | 5)/\{Pr(1) \bullet L(15 | 1)+ Pr(5) \bullet L(15 | 5)+ Pr(10) \bullet L(15 | 10)\}$$

$$= 0.2 \bullet 0.3 / \{0.3 \bullet 0.1 + 0.2 \bullet 0.3 + 0.5 \bullet 0.6\} = 0.15$$

$$c(10)= Pr(10) \bullet L(15 | 10)/\{Pr(1) \bullet L(15 | 1)+ Pr(5) \bullet L(15 | 5)+ Pr(10) \bullet L(15 | 10)$$

$$= 0.5 \bullet 0.6 / \{0.3 \bullet 0.1 + 0.2 \bullet 0.3 + 0.5 \bullet 0.6\} = 0.77$$

2.4. Histograms and Probability Density Functions

Both descriptive and inferential statistics rely on the ability to summarize probabilities, frequencies, etc, quantitatively. The central idea used to develop these summaries is the *probability density function* or PDF. The function PDF(x) expresses the probability that a field quantity has a particular numerical value, x, or a numerical value "in the vicinity of" x. The field quantity itself might take on only one of several distinct values, such as integers. The quantity, x, in this case is a *discrete variable*. An example of a discrete variable is population, since the population in a region of space can take on only integer values. If the quantity can take on any numerical value throughout a domain (e.g. in the domain [0,10]), the quantity is a *continuous variable*. An example of a continuous variable is concentration, which can have any value between 0 and infinity. Histograms are most useful in depicting discrete variables while PDFs are useful in depicting continuous variables. The same concepts may be used to describe variability and uncertainty, and so the following discussion focuses on general principles, with examples chosen from both variability and uncertainty analysis.

Imagine a field quantity such as number of individuals with a particular disease (e.g. cancer) in different grid blocks of a geographic region. The number of such individuals varies between the different grid blocks. A study is performed counting the number of such individuals in a selection of 10 grid blocks, with the data summarized in Table 2.1.

Table 2.1. Measured values of the number of individuals with cancer

Grid Block #	Measurement	$x_i - \mu$*
1	30	-9.9
2	42	+2.1
3	86	+46.1
4	18	-21.9
5	23	-16.9
6	35	-4.9
7	39	-0.9
8	46	+6.1
9	52	+12.1
10	28	-11.9

*The difference between the measured value and the mean of 39.9 (see discussion below).

How can the data in this table be summarized for descriptive statistics? Two useful measures are the mean and the variance. The general formula for the *mean*, μ, of a series of quantities is:

$$\mu = \Sigma x_i / N \tag{2.2}$$

where x_i is the numerical value of the ith field measurement and N is the total number of measures. For Table 2.1, Σx_i equals 399 and N equals 10. The mean, using Equation 2.2, then is 399/10 or 39.9. Note that the mean is simply the arithmetic average of the 10 measurements.

The mean is a measure of *central tendency*. We might also want some indication of the amount of variation in the field quantity between the grid blocks. The *variance*, Var, provides such a measure, and is calculated by the formula [6]:

$$\text{Var} = \Sigma (x_i - \mu)^2 / N \tag{2.3}$$

Note that the variance increases as the individual measurements are further from the mean. The squaring of the difference between any measurement and the mean takes place to ensure that each measurement contributes positively to the variance. Without the squaring, a measurement below the mean might cancel the contribution from a measurement above the mean. For the values in Table 2.1, the variance is:

$$\text{Var} = 3342.9 / 10 = 334.29$$

The *standard deviation* of the data, σ, equals the square root of the variance [6]:

$$\sigma = (\text{Var})^{0.5} = [\Sigma (x_i - \mu)^2 / N]^{0.5} \tag{2.4}$$

For the data in Table 2.1, where the variance is 334.29, the standard deviation is $(334.29)^{0.5}$ or 18.28 people. Note that the units of the standard deviation are the same as for the field quantity.

The standard deviation is a useful measure of variation because it will, for some PDFs, describe the fraction of samples found in certain intervals around the mean. For the normal distribution (discussed later), the interval between the mean minus the standard deviation and the mean plus the standard deviation $[\mu-\sigma,\mu+\sigma]$ contains approximately 68% of the values. For the data in Table 2.1, this interval is [39.9-18.28,39.9+18.28] or [21.62,58.18]. Note that 8 of the 10 values, or 80% of the values, in Table 2.1 fall in this range. This difference between 68% and 80% might be due to the small sample size or due to the fact that the data are not distributed normally.

It also is useful to view the data in Table 2.1 graphically. To do this, we will form a *histogram* of the data. The first step is to define distinct intervals into which the data will be grouped. For convenience, this discussion will use intervals of 10 people, defining intervals of 0 to 10, 10 to 20, etc. We then will allocate the different measurements x_i to the intervals and determine the number of measurements falling into each interval. The result is shown in Figure 2.1.

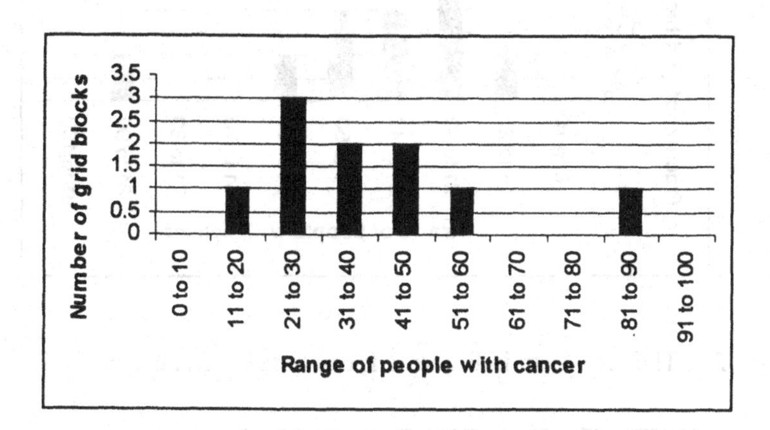

Figure 2.1. A histogram for the data in Table 2.1, showing the number of grid blocks containing a given number of people with cancer in different ranges.

A more useful formulation of the histogram for statistical analysis is to replace the number of grid blocks by the fraction of grid blocks (or frequency). This is done by dividing the number of grid blocks in a given range of the histogram by the total number of grid blocks sampled. The result for Figure 2.1 is shown in Figure 2.2.

Figures 2.1 and 2.2 were developed using data that are necessarily integer values and, hence, representative of a discrete variable. The more common problem in environmental risk assessment is to provide a statistical analysis of continuous variables. For such variables, significant information may be lost by reducing the data to a histogram.

Consider the data in Table 2.1 and the histogram in Figure 2.2. The intervals in that figure are of width 10 (i.e increments of 10 people). It would be possible to replace these 10 increments by 20 increments of width 5, or 50 increments of width 2. As the width of the increments is reduced, greater resolution is seen in the resulting histogram. If the width is reduced to an infinitesimal value, dN, the histogram reduces to a *probability density function* or PDF. The function PDF(x) indicates the fraction of the population contained in an increment "in the vicinity of" the value x. The term "in the vicinity of" means within an increment defined by [x-dx/2,x+dx/2]. Note that the total increment is equal in width to dx and centered on the value x.

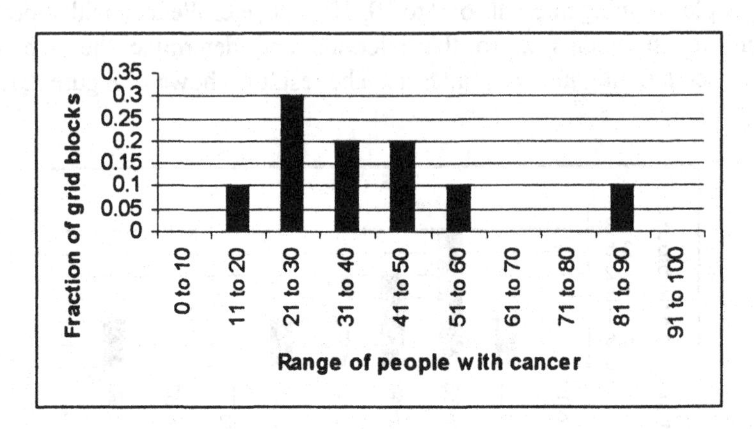

Figure 2.2. The histogram in Figure 2.1, converted to frequencies.

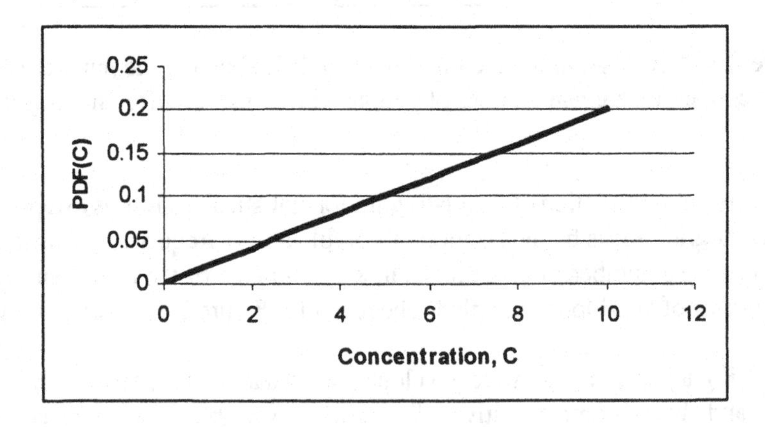

Figure 2.3. An example probability density function, showing the fraction of samples with concentration "in the vicinity of" any value, C. PDF(C) in this example increases linearly from 0 to 10, and then is 0 above a concentration of 10.

An example of a probability density function is shown in Figure 2.3. The PDF in this case is a triangular distribution with values between 0 and 10; the PDF is 0 above a concentration of 10. The height of the curve at any value of x is equal to PDF(x). For this example, the PDF is:

$$PDF(C) = 0.02C$$

for values of C between 0 and 10, and PDF(C) = 0 for all values above 10. The integral of PDF(x) from $-\infty$ to ∞ is necessarily 1, since the numerical value for every measurement must fall somewhere in this domain. In the example of Figure 2.3, since the PDF(C) equals 0.02C, the integral from 0 to ∞ (note there cannot be negative values of concentration) is:

$$\int PDF(C)\, dC = \int 0.02C\, dc = 0.02 \bullet (C)^2 / 2 = 0.02 \bullet (10)^2 / 2 = 1$$

where all integrations are from 0 to 10 in this case since the PDF is 0 for C>10.

To understand the numerical value of PDF(x), it is useful to consider a related concept, the *cumulative distribution function* or CDF(x). CDF(x) equals the fraction of samples whose value falls at or below x. Specifically, it is equal to the integral of PDF(x) from 0 to x:

$$\text{(2.5)} \qquad CDF(x) = \int PDF(x)dx$$

where the integration is from 0 to x. PDF(x) is the slope of CDF(x) evaluated at x. It expresses the rate at which CDF(x) is increasing with increasing x; where PDF(x) is high, consideration of an additional increment dx brings a greatly increased cumulative probability or frequency to CDF(x). Where PDF(x) is 0, the addition of an increment dx to CDF(x) does not increase the cumulative probability or frequency.

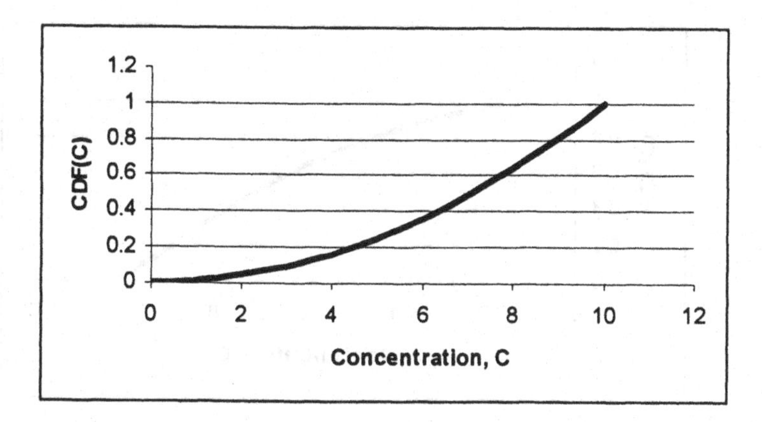

Figure 2.4. An example cumulative distribution function, showing the fraction of samples with concentration at or below any value, C. The particular function is the CDF(C) corresponding to the PDF(C) shown in Figure 2.3.

Note that CDF(x) is necessarily a monotonically increasing function for a continuous variable, beginning at 0 when x is 0 and ending at 1 for large values of x. The CDF(x) corresponding to the PDF(x) shown in Figure 2.3 may be seen in Figure 2.4. The equation describing CDF(C) for this example may be found by integrating the function PDF(C) = 0.02C from 0 to 10 (beyond 10, CDF is 1.0):

$$\text{CDF(C)} = \int \text{PDF(C)}\, dC = \int 0.02C\, dC = 0.02 \bullet C^2/2$$

where all integrations are from 0 to C.

A related concept is the *inverse cumulative distribution function* ICDF(x). ICDF(x) equals the fraction of samples whose value falls *at or above* x. Specifically, it is equal to the integral of PDF(x) from x to infinity:

$$\text{ICDF(x)} = \int \text{PDF(x)} dx \tag{2.6}$$

where the integration is from x to ∞. The ICDF(x) corresponding to the PDF(x) in Figure 2.3 and the CDF(x) in Figure 2.4 may be seen in Figure 2.5. The equation describing CDF(C) for this example may be found by integrating PDF(C) = 0.02C from 0 to 10 (beyond 10, the CDF is 1.0):

$$\text{ICDF(C)} = \int \text{PDF(C)}\, dC = \int 0.02C\, dC = (0.02 \bullet 10^2/2) - (0.02 \bullet C^2/2)$$

Where all integrations are from C to 10 since PDF is 0 for C>10.

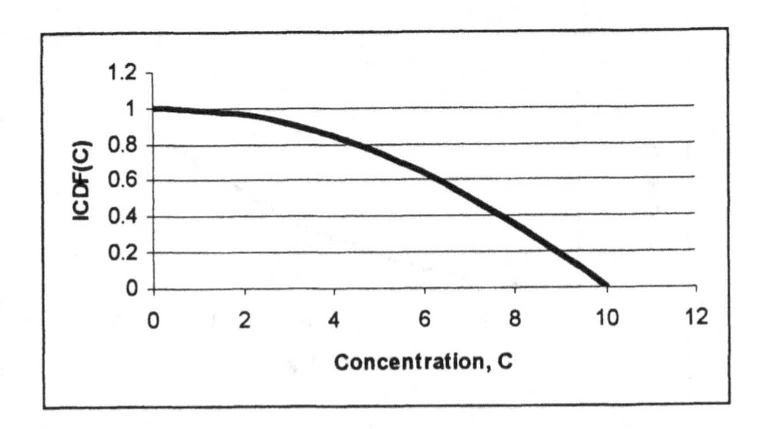

Figure 2.5. An example inverse cumulative distribution function, showing the fraction of samples with concentration at or above any value, C. The particular function is the ICDF(C) corresponding to the PDF(C) shown in Figure 2.3.

Finally, note that IDCF(x) is related directly to CDF(x) through the following equation:

(2.7) $$ICDF(x) = 1 - CDF(x)$$

2.5 Special Probability Density Functions

Several forms of the probability density function are encountered routinely in environmental risk assessment, and so these are given special treatment here. The normal distribution, $N(\mu,\sigma)$, often describes the distribution of measurement results from a measurement method applied repeatedly to a sample (e.g. a sample of soil for which the concentration is to be measured) [7]. The distribution is defined by the mean, μ, and the standard deviation, σ. An example is shown in Figure 2.6 and displays the familiar bell-shaped curve associated with the normal distribution. As mentioned previously, 68% of the values of a normally distributed quantity fall in the interval $[\mu-\sigma,\mu+\sigma]$, and 95% of the values fall in the interval $[\mu-2\sigma,\mu+2\sigma]$. The equation describing the normal PDF is:

(2.8) $$PDF(x) = \exp[(-(x-\mu)^2/(2 \bullet \sigma^2)] / [\sigma \bullet (2\pi)^{0.5}]$$

where exp is exponential function, x is the value of the field quantity at which the PDF is being evaluated, μ is the mean for the population and σ is the standard deviation for the population.

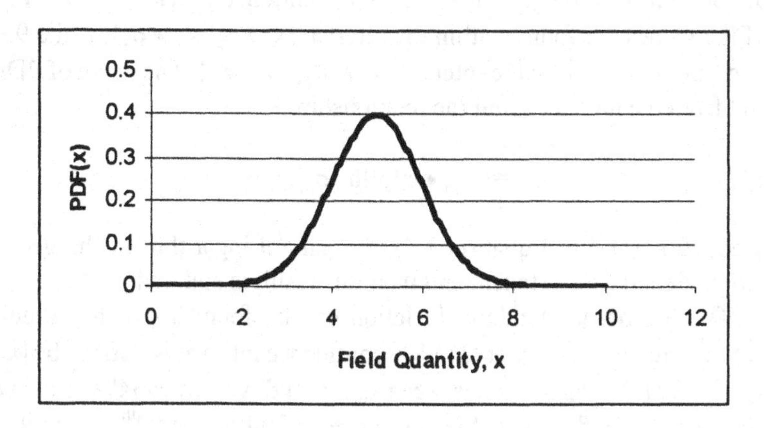

Figure 2.6. An example of a normal distribution, with $\mu = 5$ and $\sigma = 1$.

A wide variety of environmental and biological quantities encountered routinely in risk assessment are characterized by variability described by the *lognormal distribution* [8] For such quantities, the natural logarithm of the numerical values for the samples are distributed normally. It should be evident from Figure 2.6 that the normal distribution is *symmetrical*; i.e. the PDF has an equal height at values of the quantity equally far removed from the mean in both directions. By contrast, the lognormal distribution is *asymmetrical* or *skewed*.

The PDF for the lognormal distribution is shown as $LN(x_m,\sigma_g)$, where x_m is the *median* of the distribution and σ_g is the *geometric standard deviation*. The median is the value of the field quantity lying at the 50th percentile of the cumulative distribution function (i.e. 50% of the values in the distribution are below the median and 50% are above). For a normal distribution, the median and the mean are identical, since the distribution is symmetrical. For the lognormal distribution, the median is less than the mean. In addition, consider the *mode* of a distribution, which is the value of the field quantity for which the PDF is at a maximum. The mode, median and mean are identical in a normal distribution, but not for a lognormal distribution. For a lognormal distribution, the mode is less than the median, which in turn is less than the mean.

The function describing PDF(x) for the lognormal distribution is:

$$(2.9) \quad PDF(x) = \exp[(-(\ln x - \ln x_m)^2/(2 \bullet \ln^2 \sigma_g)] / [x \bullet \sigma_g \bullet (2\pi)^{0.5}]$$

Figure 2.7 provides an example of a lognormal PDF. The confidence intervals for the lognormal distribution are developed in the same manner as for a normal distribution, but with the log transformation mentioned previously. Specifically, 68% of the values are found within the interval $[x_m / \sigma_g; x_m \bullet \sigma_g]$; while 95% of the values are found within the interval $[x_m / \sigma^2_g; x_m \bullet \sigma^2_g]$. The mean of PDF can be found from the median using the relationship:

$$(2.10) \quad \mu = x_m \bullet \exp(\ln^2(\sigma_g)/2)$$

In this equation (and in Equation 2.9), the natural logarithm of the geometric standard deviation first is taken, and then this result is squared.

The geometric standard deviation can be found from the cumulative distribution function. Note that the 68% confidence interval is defined by $[x_m / \sigma_g; x_m \bullet \sigma_g]$. 34% of the values are between x_m / σ_g and x_m, while 34% of the values found between x_m and $x_m \bullet \sigma_g$. Since x_m is by definition the 50th percentile, $x_m \bullet \sigma_g$ must lie at the 84th percentile (50% + 34%) and x_m / σ_g must lie at the 16th percentile (50% - 34%).

Example 2.3. A normal distribution is characterized by a mean, μ, of 2.5 and a standard deviation, σ, of 1.1. What is the interval, constructed symmetrically about the mean, containing 68% of the values?

$$[\mu-\sigma,\mu+\sigma] = [2.5-1.1,2.5+1.1] = [1.4,3.6]$$

The interval containing 95% of the values, again constructed symmetrically about the mean, is:

$$[\mu-2\sigma,\mu+2\sigma] = [2.5-2\bullet 1.1,2.5+2\bullet 1.1] = [2.5-2.2,2.5+2.2] = [0.3,4.7]$$

Now consider a lognormal distribution with mean, μ, of 2.5 and geometric standard deviation, σ_g, of 2. What is the interval, constructed about the median, containing 68% of the values? First, the mean must be converted to a median using Equation 2.10:

$$\mu = x_m \bullet \exp(\ln^2(\sigma_g)/2) \quad \text{or} \quad x_m = \mu \bullet \exp(-\ln^2(\sigma_g)/2)$$

Therefore, $x_m = 2.5 \bullet \exp(-\ln^2(2)/2) = 1.97$

The interval containing 68% of the values is:

$$[x_m / \sigma_g; x_m \bullet \sigma_g] = [1.97/2,1.97\bullet 2] = [0.99.3.94]$$

The interval containing 95% of the values is:

$$[x_m / \sigma^2_g; x_m \bullet \sigma^2_g] = [1.97/4,1.97\bullet 4] = [0.49.7.89]$$

Both the normal and lognormal distributions are continuous. In some cases, field quantities may be discrete, as in the field of the number of people with effects in different grid blocks, and stochastic. Two discrete distributions applicable to stochastic processes, which are related, appear quite often in risk assessment: the *binomial* and *Poisson*.

The binomial distribution describes quantities in which there is a population of individuals, each of which has some probability of showing the effect of interest (perhaps cancer). Let N equal the total number of individuals in the population and P_i equal the probability that a randomly selected individual will develop the effect. The probability that exactly R individuals in this population will show the effect during a single measurement is:

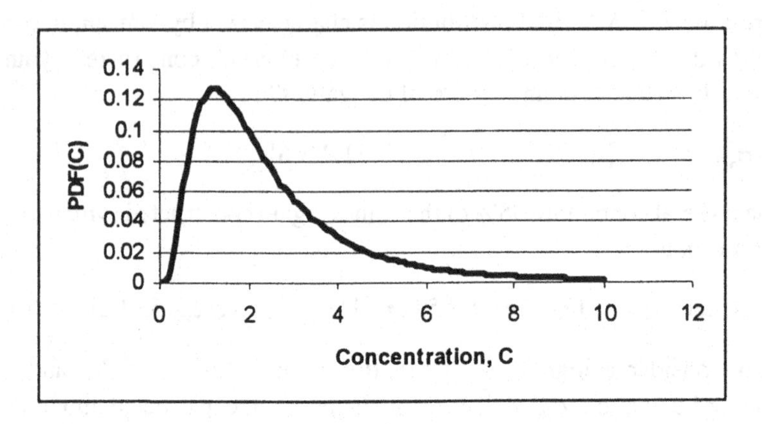

Figure 2.7. An example of a lognormal distribution, with median of 2 and geometric standard deviation of 2. Note the skewing of the distribution to the right. The mode is less than the median, and the mean is greater than the median.

(2.11) $$P(R) = N! \bullet P_i^R \bullet (1-P_i)^{N-R} \,/\, [R! \bullet (N-R)!]$$

Example 2.4. Consider a case in which each individual has a probability P_i of 0.01 of developing a rash. There are 30 individuals in the population (N is 30). What is the probability that 6 individuals develop the rash? Using Equation 2.11:

$$P(6) = 30! \bullet 0.01^6 \bullet (1-0.01)^{30-6} / [6! \bullet (30-6)!] = 4.7 \bullet 10^{-7}$$

The Poisson distribution is a special case of the binomial distribution, useful when N in Equation 2.11 is large and P_i is small. If N is the population size and Pi is the probability of a randomly selected individual in the population developing an effect, the mean number of effects can be calculated from:

(2.12a) $$\mu = N \bullet P_i$$

The probability that exactly R individuals in the population develop the effect then is:

(2.13) $$P(R) = \mu^R \bullet \exp(-\mu) / R! = (N \bullet P_i)^R \bullet \exp(-N \bullet P_i) / R!$$

Example 2.5. Consider the same problem as in Example, 2.4. Each individual has a probability P_i of 0.01 of developing a rash. There are 30 individuals in the population (N is 30). What is the probability that 6 individuals develop the rash, assuming the Poisson approximation is valid? Using Equation 2.12:

$$\mu = N \bullet P_i = 30 \bullet 0.01 = 0.3$$

Using Equation 2.13:

$$P(6) = \mu^R \bullet \exp(-\mu) / R! = (0.3)^6 \bullet \exp(-0.3) / 6! = 7.5 \bullet 10^{-5}$$

Note that this answer is approximately a factor of 1.6 higher than the answer in Example 2.4 using the binomial distribution. This is due to the fact that the Poisson distribution is only an approximation, requires a large value of N (even an N of 30 is rather small), and improves as P_i is reduced (even 0.01 is rather high).

Cases arise in risk assessment where the parametric form of a distribution is unknown (i.e. whether it is normal, lognormal, etc) and where there is little information on which to fully specify the distribution. At such times, it may be necessary to use less detailed distributions. Suppose, for example, that one has a best estimate of a parameter value as well as upper and lower limits. A possibility is to construct a triangular PDF with the following characteristics:

- The PDF is 0 below the lower limit, X_L
- The PDF rises linearly from 0 at X_L to a peak at the best estimate, X_B
- The PDF decreases linearly from X_B to the upper limit, X_U
- The PDF is 0 above the upper limit, X_U

PDF(x) then may be described by the relations:

(2.14) If $x < X_L$, $PDF(x) = 0$

(2.15) If $X_L < x < X_B$, $PDF(x) = P(X_B) \bullet (x - X_L) / (X_B - X_L)$

(2.16) If $X_B < x < X_U$, $PDF(x) = P(X_B) - P(X_B) \bullet (x - X_B) / (X_U - X_B)$

(2.17) If $x > X_U$, $PDF(x) = 0$

and where:

(2.18) $P(X_B) = 2 / (X_U - X_L)$

At times, even less information is available to develop the PDF. Perhaps only the upper and lower limits can be estimated. At such times, it is common to use a uniform PDF, with equal values of the PDF between the lower limit, X_L, and the upper limit, X_U. The resulting PDF is described by the relations:

(2.19) If $x < X_L$, $PDF(x) = 0$

(2.20) If $X_L < x < X_U$, $PDF(x) = 1 / (X_U - X_L)$

(2.21) If $x > X_U$, $PDF(x) = 0$

2.6. Correlation

At times, two distributed field quantities may be *correlated*. Knowing the first value provides information about the second (if this is not true, the two quantities are *independent*). If a randomly selected value of the first field quantity is above the mean of its underlying PDF, the quantity from the second field may also be above its mean. Similarly, if a randomly selected value of the first field quantity is below the mean of its underlying PDF, the quantity from the second field may also be below its mean. If this pattern holds true in general for these two fields, there is a *positive correlation* between the two field quantities. If the inverse pattern (a value above the mean for the first quantity is associated with a value below the mean for the second quantity) holds true in general for these two fields, there is a *negative correlation* between the two field quantities.

At least the direction of a correlation (positive or negative) can be seen from examination of a scatter plot. Consider two field quantities, x and y (perhaps temperature and humidity). Pairs of measurements are made in which each of the quantities is measured at the same point in space and the same time, summarized as (x_i,y_i). A scatter plot of these two quantities depicts the plane x-y, with each pair indicated by a point on the plane. An example is shown in Figure 2.8. The figure on the left is an example of positive correlation since, in general, a higher value of x corresponds to a higher value of y. The figure on the right is an example of negative correlation since, in general, a higher value of x corresponds to a lower value of y.

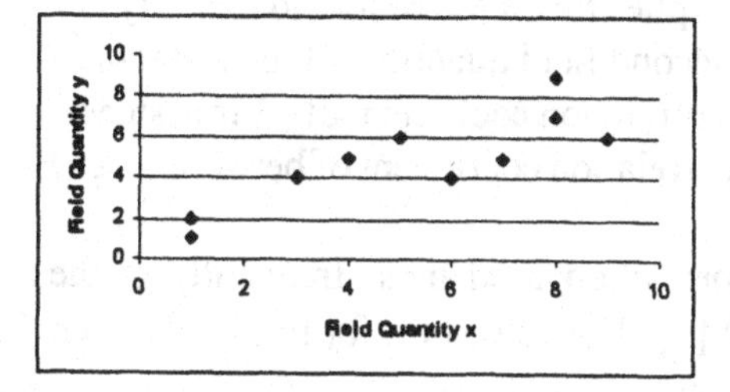

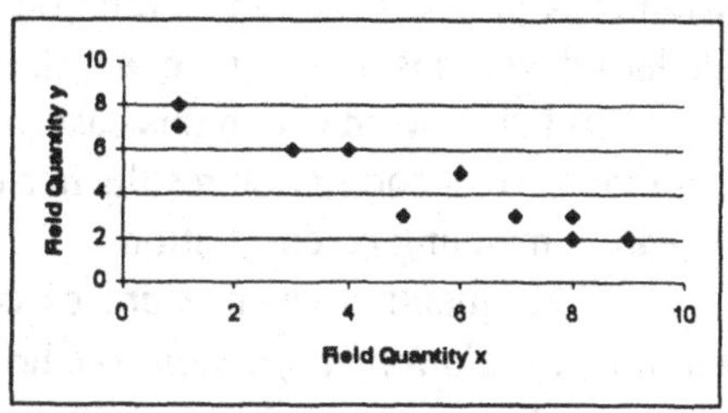

Figure 2.8. Scatter plots for two positively correlated field quantities (left figure) and for two negatively correlated field quantities (right figure).

In addition to the direction of the correlation, it is useful to develop a measure of the extent or magnitude of the correlation. The more strongly correlated two quantities, the greater the degree to which they form a smooth line. The two quantities in Figure 2.8 (either of the two scatter plots) have a fairly strong correlation, although evidently not perfect since the points do not fall on a smooth line.

To develop a quantitative measure of correlation, it is useful to consider re-formatting the data used in a scatter plot. Instead of the numerical values for each quantity, a *z-score* is calculated [2]. The z-score, also referred to as a *standard score*, is defined as the difference between a particular field quantity's value and the mean for the underlying distribution from which it was drawn, divided by the standard deviation for that distribution:

(2.22) $$z_i = (x_i - \mu) / \sigma$$

The value of z is positive if the measurement is above the mean, and negative if it is below the mean. A given deviation from the mean is more important (i.e. z is higher) if the standard deviation is small and, hence, it is less likely to find such a large deviation.

The advantage of the z-score over the original measurement value is that it normalizes or *standardizes* the data from measurements of two different quantities. In particular it changes the scale of the measurements into a unit that represents the number of standard deviations of the measurement result from the mean of the underlying distribution. If two quantities are perfectly correlated, they will lie at the same number of standard deviations from their associated mean. A perfect positive correlation implies that if the first field quantity is z standard deviations above its mean, the second field quantity also will be z standard deviations above its own mean. In this case, a correlation coefficient of +1 is assigned. A perfect negative correlation implies that if the first field quantity is z standard deviations above its mean, the second field quantity will be z standard deviations below its mean. In this case, a correlation coefficient of −1 is assigned. Less than perfect correlation results in a correlation coefficient of between −1 and +1, with 0 meaning no correlation.

The quantitative measure of correlation used most frequently is the *Pearson correlation coefficient*, ρ (rho) [2]. The value of ρ for two distributed field quantities is:

$$\rho = \Sigma z_1 \bullet z_2 / N \tag{2.23}$$

where z_1 is the z-score for a measurement from the first field, z_2 is the z-score for the corresponding measurement from the second field, and N is the number of pairs of measurements available. The summation is over all N pairs.

Example 2.6. Consider the data underlying Figure 2.8 (left figure). The 10 pairs of measurements for the two field quantities are (1,2), (5,6), (3,4), (9,6), (7,5), (6,4), (1,1), (8,7), (8,9) and (4,5). The mean of the first quantity is 5.2 and the standard deviation is 2.89. The mean of the second quantity is 4.9 and the standard deviation is 2.33. The z-scores for the 10 measurements of the first quantity then are 1.5, 0.1, 0.8, -1.3, -0.6, -0.3, 1.5, -1.0, -1.0, 0.4. The z-scores for the 10 measurements of the second quantity then are 1.2, -0.5, 0.4, -0.5, -0.05, 0.4, 1.7, -0.9, -1.8, -0.04. Using Equation 2.15, the Pearson correlation coefficient is:

$$\rho = \Sigma z_1 \bullet z_2 / N = [(1.5 \bullet 1.2) + (0.1 \bullet -0.5) + (0.8 \bullet 0.4) + (-1.3 \bullet -0.5)$$

$$+ (-0.6 \bullet -0.05) + (-0.3 \bullet 0.4) + (1.5 \bullet 1.7) + (-1.0 \bullet -0.9) + (-1.0 \bullet -1.8)$$

$$+ (0.4 \bullet -0.04)] / 10 = 0.76$$

As mentioned previously, this is a moderate positive correlation between the two field quantities.

2.7. Parameter Estimation and Measures of Model Quality

The discussion in this chapter has focused on summarizing the statistical properties of a set of data, and comparing two sets of data to test for correlation. We turn now to the issue of using data to develop best-fitting parameter values for a model. This section reviews the fundamentals of two methods: *least-squares* and *maximum likelihood* [9]. Each compares competing parameter estimates, and their associated model fits to the data, through development of a measure of *goodness of fit*.

Under empirical tests of a model, in which the success of a model is measured solely by the ability to fit a set of data, a best-fitting model is one that in some sense has the smallest overall difference between the data points and model predictions at those same points. In the *least-squares method*, this difference is literally the arithmetic difference between all pairs of model predictions and data points. Consider a set of pairs of results (d_i, m_i) where d_i is the ith data point and m_i is the model prediction under conditions identical to those when d_i was collected. The difference between the ith data point and model prediction is $(d_i - m_i)$. This is shown graphically in Figure 2.9. As a simple measure of goodness of fit, one might add these differences as follows:

$$(2.24) \qquad GOF = \Sigma (d_i - m_i)^2$$

where GOF is the goodness of fit measure, the summation is over all N data points and their associated model predictions, and the difference $(d_i - m_i)$ is squared so positive and negative differences do not cancel (i.e. it is equally bad for the model to over as to under-predict).

Equation 2.24 is applied by searching for the value of all parameters (parameters appearing in the model that yields the values of m_i) such that GOF is minimized (hence the term least squares). Perfect agreement between the data and the model predictions produces a GOF measure that is 0. The process of finding these best-fitting parameter values may be either *analytic* (through derivation) or *procedural* (through systematic or random searches for the parameter values). The analytic approach uses the concept of *minima* and *maxima* in functions.

Consider the function $F(x) = cx$. There are N data points available, shown as $d_1, d_2 \ldots d_N$. For each of these, there is a corresponding value of x, shown as x_1, $x_2 \ldots x_N$. From Equation 2.24:

$$(2.25) \qquad GOF = \Sigma (d_i - m_i)^2 = (d_1 - cx_1)^2 + (d_2 - cx_2)^2 + \ldots (d_N - cx_N)^2$$

Example 2.7. Consider a model of the form F(x) = cx, where c is a parameter (constant with x). The task is to determine the value of c. Three data points are available. At x=0, F(x) is measured to be 0.3. At x=5, F(x) is measured to be 0.9. At x=30, F(x) is measured to be 5.3. What is the goodness of fit measure, using Equation 2.24, if c is 0.2?

$$GOF = \Sigma (d_i - m_i)^2 = (0.3\text{-}0.2\bullet 0)^2 + (0.9\text{-}0.2\bullet 5)^2 + (5.3\text{-}0.2\bullet 30)^2 = 0.59$$

What is the goodness of fit measure if c is 0.4?

$$GOF = \Sigma (d_i - m_i)^2 = (0.3\text{-}0.4\bullet 0)^2 + (0.9\text{-}0.4\bullet 5)^2 + (5.3\text{-}0.4\bullet 30)^2 = 46.2$$

In this example, the parameter value of 0.2 for c provides a better fit of the model to the data, since the GOF measure is better than that for c equal to 0.4.

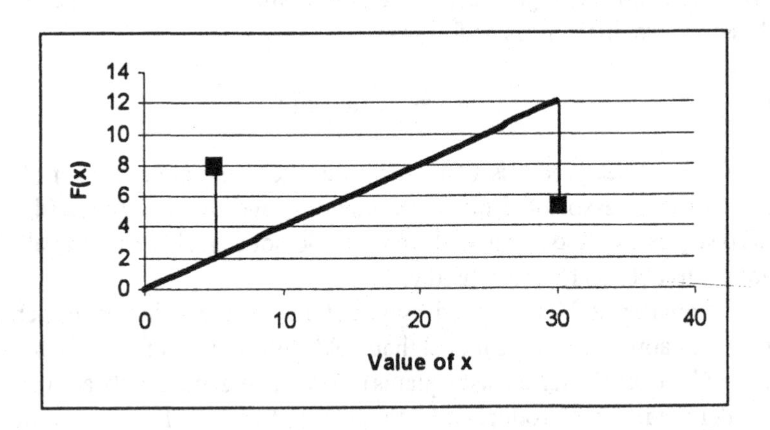

Figure 2.9. A comparison of two data points against a model. The vertical distances between the data and the predictions are shown as the two vertical lines. These distances equal the values of $(d_i\text{-}m_i)$ shown in Equation 2.24. A better fit corresponds to a smaller value of these vertical differences.

As c changes, so does GOF. For some value of c, GOF will be minimized. This is the value of c corresponding to the least squares. This also will be the value of c for which the slope of GOF versus c is 0. Taking the derivative of both sides of Equation 2.25 and setting it equal to 0:

(2.26a) $dGOF/dc = -2\,(d_1\text{-}cx_1)x_1 - 2\,(d_2\text{-}cx_2)x_2 - \ldots.\ 2\,(d_N\text{-}cx_N)x_N = 0$

Dividing both sides by –2:

(2.26b) $(d_1\text{-}cx_1)x_1 + (d_2\text{-}cx_2)x_2 + \ldots.\ (d_N\text{-}cx_N)x_N = 0$

Rearranging terms:

(2.26c) $(d_1x_1 - cx^2_1) + (d_2x_2 - cx^2_2) + \ldots.\ (d_Nx_N - cx^2_N) = 0$

or

(2.26d) $cx^2_1 + cx^2_2 + \ldots.\ cx^2_N = d_1x_1 + d_2x_2 + \ldots\ d_Nx_N$

or

(2.26e) $c = [d_1x_1 + d_2x_2 + \ldots\ d_Nx_N] / [x^2_1 + x^2_2 + \ldots.\ x^2_N]$

Since the (d_i,x_i) pairs all are known, it is possible to solve for c. The result is the value of c corresponding to the least squares. Note that if there is a single data point, c equals d_1/x_1, or the curve passes directly through that one data point as expected. In general, the process for finding the best-fitting parameter is to find the value of the parameter satisfying:

(2.27) $dGOF/dc = 0$

Equation 2.24 assigns each data point equal weight in developing a measure of goodness of fit. In practice, there may be reasons why the ability of the model to "pass near" some data points may be given higher weight. For example, the latter data points may be closer to the part of the domain in which predictions ultimately are needed in a risk assessment, or those data may be more reliable. These possibilities are reflected in use of a weighted least squares method:

(2.28) $GOF_w = \Sigma\,(d_i - m_i)^2 \bullet W_i$

where the subscript w stands for "weighted" and where W_i is the weighting factor for the ith data point. One commonly used weighting factor recognizes that data points that are highly uncertain should be counted less than other data points. One measure of uncertainty is the standard deviation divided by the mean (the standard error), σ_i / d_i. The higher this value for a data point, the lower the weighting factor, so Equation 2.28 in this case becomes:

(2.29) $$GOF_w = \Sigma (d_i - m_i)^2 / (\sigma_i / d_i)$$

Example 2.8. Consider the model and data in Example 2.7. The standard deviation for the data point at x=0 is 0.1; the standard deviation for the data point at x=5 is 0.8; the standard deviation for the data point at x=30 is 1. What is the goodness of fit measure, using Equation 2.29, if c is 0.2?

$$GOF = \Sigma (d_i - m_i)^2 = (0.3-0.2 \bullet 0)^2 / (0.1/0.3) + (0.9-0.2 \bullet 5)^2 / (0.8/0.9) + (5.3-0.2 \bullet 30)^2 / (1/5.3) = 2.87$$

Generally, the best-fitting parameter value determined by the unweighted and weighted least squares procedures will differ.

The method of *maximum likelihood* uses the methods of probability theory more directly to locate best-fitting parameter values. Rather than using the vertical distance between data and predictions (a geometrical measure), the maximum likelihood uses a measure based on the likelihood that the data would have been produced if the model predictions were correct. For a single data point, the measure of goodness of fit then is:

(2.30) $$GOF_i = L(d_i | m_i)$$

If there are N data points, the likelihood of obtaining all N of the specific measured values will equal the product of the individual likelihoods:

(2.31) $$GOF = \Pi L(d_i | m_i)$$

where Π represents the product over all values i.

For example, the Poisson distribution often is used to calculate the likelihoods when the measurement results must be discrete values (e.g. the number of deaths in a population). Referring back to Equation 2.22:

(2.32) $$L(d_i | m_i) = \mu^R \bullet \exp(-\mu) / R! = m_i^{di} \bullet \exp(-m_i) / d_i!$$

where R in Equation 2.32 is now the measurement result, d_i, and μ in that equation is now the model prediction, m_i.

Example 2.9. Consider a model and two data points. The data points are measurements of the number of cases of respiratory disease in a population and the model is F(C) = bC, where C is the air concentration of a pollutant in g/m^3, b is a constant, and F(C) is the probability that a randomly selected individual shows the effect. The size of the exposed population is 1000; therefore, the mean number of effects is b • C • 1000. When C is 2 g/m^3, a measurement yields 5 people showing the effect. When C is 10 g/m^3, 14 people show the effect. What is the goodness of fit measure, using Equations 2.31 and 2.32, if b is 0.002?

$$GOF = \Pi L(d_i | m_i) = \Pi m_i^{d_i} \bullet \exp(-m_i) / d_i!$$

$$= [(0.002\bullet 2\bullet 1000)^5 \bullet \exp(-0.002\bullet 2\bullet 1000) / 5!] \bullet [(0.002\bullet 10\bullet 1000)^{14} \bullet$$

$$\exp(-0.002\bullet 10\bullet 1000) / 14!] = 0.156 \bullet 0.039 = 0.006$$

The procedure for finding the maximum likelihood estimate of the parameter value proceeds as in the case of the least squares method, using either analytic (for very simple models) or sampling methods. The sole difference is that in the least squares method, the search is for the lowest sum of the squares, while for the maximum likelihood method, the search is for the largest product of the likelihoods.

Both the least squares and maximum likelihood methods usually require a search for the parameter value with the best goodness of fit measure (assuming an analytic approach is not available). How can this search be conducted most effectively? This issue is particularly important if a model has many parameters ($x_1, x_2, \ldots x_N$) and the task is to find the best fitting set of these parameters. Three possibilities are used routinely in risk assessment:

- *Gridded search*. In this approach, the domain for each of the parameters is divided into a grid of discrete points. For example, a parameter might be known to lie somewhere between 0 and 100. This domain could be divided into increments of 10, and values of the parameter equal to 0, 10, 20, ... 100 examined for the GOF measure. All combinations of these discrete values of the different parameters are considered by calculating a GOF measure for each and selecting the set with the best value of the GOF measure. The problem with such an approach is that it may result in a large number of parameter values to consider if the domain is large and/or there are many different parameters in the model (each requiring their own grids). In addition, it is possible that a best-fitting set of parameter values obtained from such a procedure will miss an even better fitting set of values due to the resolution of the grid.

- *Random search*. In this approach, values of the set (x_1, x_2, ... x_N) of parameters are generated at random (see the discussion on Monte Carlo modeling in Chapter 7). Each randomly selected set is used to calculate a GOF measure, and the set with the best GOF measure selected. This approach avoids the problem of using only the values from a pre-defined grid, but it still examines only a finite set of values for (x_1, x_2, ... x_N), and the possibility remains that a better fitting set of values exists.

- *Method of descent*. In this approach, an initial set of parameter values (x_1, x_2, ... x_N) is introduced. The GOF measure is calculated for this set. An incremental change in a parameter value is established (e.g. Δx_1). The parameter value from the first set is adjusted upwards by this amount and a new GOF measure calculated. If the new GOF measure is better than the first, another increment Δx_1 is added to x_1 (producing $x_1 + 2\Delta x_1$) and a new GOF calculated. If this third GOF is better than the second, the process continues in the same direction, with subsequent additions of Δx_1. Eventually, the GOF value for iteration n is not better than the GOF value for iteration n-1. If this occurs, either the procedure moves backwards (i.e. by subtracting values of Δx_1) or one moves back to the n-1 iteration and tries adding a smaller increment such as $\Delta x_1/2$. The procedure continues, with progressively smaller values of Δx_1, until the GOF measure ceases to increase, at which point a best fitting value for the parameter has been located. To increase the resolution, Δx_1 is decreased. The problem with such an approach is that one might "walk" into a local region of the domain in which some value of the parameter produces an apparently best GOF measure, but there may be a "valley" or "peak" (depending on whether a least squares of maximum likelihood approach is being used as the GOF measure) somewhere else in the domain. To avoid this, it will be necessary to repeat the process using randomly selected starting sets of parameter values. In addition, when there is more than one parameter value, it is inefficient to adjust one parameter value at a time, and may cause large errors in finding the set of parameter values with the overall best GOF measure. In this case, a better approach is to begin with an initial set of parameter values, calculate the gradient (see Chapter 1), and move in the direction of steepest descent of the gradient. This method is beyond the scope of this book.

In Examples 2.10 through 2.12 shown below, the problem is sufficiently simple to allow use of the analytic method. Returning to Equation 2.26e:

$$c = [d_1x_1 + d_2x_2] / [x^2_1 + x^2_2] = [3 \bullet 5 + 7 \bullet 10] / [5^2 + 10^2] = 0.68$$

For the 3 examples, the best values of c were 0, 0.6 and 0.5, respectively. The differences between these answers and the true best fitting value of 0.68 is simply due to the resolution applied in the examples. In each case, this resolution could have been increased and values much closer to 0.68 obtained. While the answer will never be exactly 0.68 (unless luck intervenes!), one can get an answer using any of the three approaches which is arbitrarily close by improving the grid resolution (for Example 2.10), selecting more random numbers (for Example 2.11), or continuing the process of iteration further (for Example 2.12).

Example 2.10. Consider a simple model of the form F(x) = cx, where c is a constant to be determined. There are two data points. When x is 5, F(x) is measured to be 3. When x is 10, F(x) is measured to be 7. Using an unweighted least-squares criterion, what is the best-fitting value for c using the gridded domain approach?

Note that c is probably somewhere between 0 and 10. We will use a grid for c between 0 and 10 with increments of 5 (just to make this example *very* simple). The value of c may, therefore, be 0, 5 or 10. Starting with the value of 0, and using Equation 2.24, the GOF measure is:

$$GOF = \Sigma (d_i - m_i)^2 = (3 - 0 \bullet 5)^2 + (7 - 0 \bullet 10)^2 = 58$$

For c equal to 5:

$$GOF = (3 - 5 \bullet 5)^2 + (7 - 5 \bullet 10)^2 = 2333$$

And for c equal to 10:

$$GOF = (3 - 10 \bullet 5)^2 + (7 - 10 \bullet 10)^2 = 10858$$

Therefore, the best value for c is 0 by this approach. A better value might be found by selecting a smaller grid resolution.

Example 2.11. Consider a simple model of the form $F(x) = cx$, where c is a constant to be determined. There are two data points. When x is 5, F(x) is measured to be 3. When x is 10, F(x) is measured to be 7. Using an unweighted least-squares criterion, what is the best-fitting value for c using a random sampling approach?

Note that c is probably somewhere between 0 and 10. We will use values for c selected randomly between 0 and 10. These random values are 6.4, 0.6 and 4.2. Starting with the value of 6.4, and using Equation 2.24, the GOF measure is:

$$GOF = \Sigma (d_i - m_i)^2 = (3 - 6.4 \bullet 5)^2 + (7 - 6.4 \bullet 10)^2 = 4090$$

For c equal to 0.6:

$$GOF = (3 - 0.6 \bullet 5)^2 + (7 - 0.6 \bullet 10)^2 = 1$$

And for c equal to 4.2:

$$GOF = (3 - 4.2 \bullet 5)^2 + (7 - 4.2 \bullet 10)^2 = 1549$$

Therefore, the best value for c is 0.6 by this approach. A better value might be found by taking a larger random sample.

2.8. Error Propagation through Models

While the uncertainty or variability associated with each parameter in a field equation might be known, one is usually interested in the overall uncertainty or variability in the prediction resulting from the equation. This requires a method for *propagating* the uncertainty or variability through the equation, or estimating the variance of the predictions conditional on the variance of the parameters appearing within the defining equation. The general relationship for finding the overall variance is [10]:

$$(2.33) \quad Var(F(c_1,c_2)) = Var_{c1}[dF(c_1,c_2)/dc_1]^2 + Var_{c2}[dF(c_1,c_2)/dc_2]^2$$

Example 2.12. Consider a simple model of the form F(x) = cx, where c is a constant to be determined. There are two data points. When x is 5, F(x) is measured to be 3. When x is 10, F(x) is measured to be 7. Using an unweighted least-squares criterion, what is the best-fitting value for c using a descent approach?

We will start with c equal to 0 and "walk" in the positive c direction (arguing that c cannot be negative). The increment of Δc will be 1 initially. Beginning with c equal to 0:

$$\text{GOF} = (3-0 \bullet 5)^2 + (7-0 \bullet 10)^2 = 58$$

Moving to c equal to 1:

$$\text{GOF} = (3-1 \bullet 5)^2 + (7-1 \bullet 10)^2 = 13$$

Since 13 is better than 58 for a least squares method, we will continue moving in this direction. Trying c equal to 2 $(1 + \Delta c)$:

$$\text{GOF} = (3-2 \bullet 5)^2 + (7-2 \bullet 10)^2 = 218$$

218 is worse than 13, so we have "shot past" the optimal solution. Return to c equal to 1 and reduce Δc to 0.5. The new value of c is 1.5:

$$\text{GOF} = (3-1.5 \bullet 5)^2 + (7-1.5 \bullet 10)^2 = 84$$

which is still worse than 13. So, try moving backwards from 1 by subtracting Δc from 1 to get c equal to 0.5:

$$\text{GOF} = (3-0.5 \bullet 5)^2 + (7-0.5 \bullet 10)^2 = 4.3$$

Which is better than 13. Now, continue around the point c equal to 0.5 with an even smaller Δc (perhaps 0.1) until the best value of c is obtained with some desired resolution. Note that at the moment, the resolution is 0.5, since all we can say is that the best value of c is within 0.5 of the value 0.5.

The value of 0.5 is a "local optimum". It resulted from movement from c equal to 0, to c equal to 1, to c equal to 2, and back to c equal to 0.5. We cannot be sure, however, if a better fitting value of c might have existed past c equal to 2. To resolve this problem, and to find a "global optimum (a solution that is best for the entire domain) we might start the process from a new point, such as c equal 5.

where $Var(F(c_1,c_2))$ is the variance of the function F containing the parameters c_1 and c_2; Var_{c1} is the variance of the parameter c_1; Var_{c2} is the variance of the parameter c_2; $dF(c_1,c_2)/dc_1$ is the partial derivative of the function with respect to c_1; and $dF(c_1,c_2)/dc_2$ is the partial derivative of the function with respect to c_2. The partial derivatives are evaluated at the mean values of all terms appearing in the derivative. If there are more parameters, the sum in Equation 2.33 continues through all N parameters. As examples, consider several simple functions below:

2.5.1. $F(x) = c_1x$

Using Equation 2.33:

$$(2.34) \qquad Var(F) = Var_{c1}[dF/dc_1]^2 = Var_{c1}[x^2]$$

Example 2.13. Let F(x) = cx. The mean of c is 0.2 and the variance of c is 0.05. What is the variance of F(x) for x equal 4? Using Equation 2.34:

$$Var(F) = Var_{c1}[x^2] = 0.05 \bullet 4^2 = 0.8$$

The mean estimate of F at x equal 4 is $0.05 \bullet 4 = 0.2$

2.5.2. $F(x) = (c_1 + c_2)x$

Using Equation 2.33:

$$(2.35) \quad Var(F) = Var_{c1}[dF/dc_1]^2 + Var_{c2}[dF/dc_2]^2 = Var_{c1}[x]^2 + Var_{c2}[x]^2$$

Example 2.14. Let $F(x) = (c_1 + c_2)x$. The mean of c_1 is 0.2 and the variance of c_1 is 0.05. The mean of c_2 is 0.9 and the variance of c_2 is 0.3. What is the variance of F(x) for x equal 4? Using Equation 2.35:

$$Var(F) = Var_{c1}[x]^2 + Var_{c2}[x]^2 = 0.05 \bullet 4^2 + 0.3 \bullet 4^2 = 5.6$$

The mean estimate of F at x equal 4 is $(0.2+0.9) \bullet 4 = 4.4$

2.5.3. $F(x) = c_1c_2x$

Using Equation 2.33:

(2.36) $$Var(F) = Var_{c1}[dF/dc_1]^2 + Var_{c2}[dF/dc_2]^2 = Var_{c1}[c_2x]^2 + Var_{c2}[c_1x]^2$$

Example 2.15. Let $F(x,y) = c_1c_2x$. The mean of c_1 is 0.2 and the variance of c_1 is 0.05. The mean of c_2 is 0.9 and the variance of c_2 is 0.3. What is the variance of F(x) for x equal 4? Using Equation 2.36:

$$Var(F) = Var_{c1}[c_2x]^2 + Var_{c2}[c_1x]^2 = 0.05 \bullet [0.9 \bullet 4]^2 + 0.3 \bullet [0.2 \bullet 4]^2 = 0.84$$

The mean estimate of F at x equal 4 is $0.2 \bullet 0.9 \bullet 4 = 0.72$

2.5.4. $F(x) = c_1x/c_2$

Using Equation 2.33:

(2.37) $$Var(F) = Var_{c1}[dF/dc_1]^2 + Var_{c2}[dF/dc_2]^2 = Var_{c1}[x/c_2]^2 + Var_{c2}[-x/c_1]^2$$

Note: this derivation used the following expression:

(2.38) $$d(A/B)/dc = (B \bullet dA/dc - A \bullet dB/dc) / B^2$$

Example 2.16. Let $F(x,y) = c_1x/c_2$. The mean of c_1 is 0.2 and the variance of c_1 is 0.05. The mean of c_2 is 0.9 and the variance of c_2 is 0.3. What is the variance of F(x) for x equal 4? Using Equation 2.37:

$$Var(F) = Var_{c1}[x/c_2]^2 + Var_{c2}[x/c_1]^2 = 0.05 \bullet [4/0.9]^2 + 0.3 \bullet [4/0.2]^2 = 121$$

The mean estimate of F at x equal 4 is $0.2 \bullet 4 / 0.9 = 0.89$

Note how large the variance of a ratio can be.

References

1. T. Wonnacott and R. Wonnacott, *Introductory Statistics*, John Wiley and Sons, New York, 1977.
2. J. Fruend and G. Simon, *Statistics: A First Course*, Prentice Hall, New Jersey, 1995.
3. R. Burford, *Introduction to Finite Probability*, Charles E. Merril Books, Inc., Columbus, Ohio, 1967.
4. S. Silvey, *Statistical Inference*, Penguin Books, Baltimore, 1970.
5. R. Jeffrey, *The Logic of Decision*, University of Chicago Press, Chicago, 1983.
6. S. Kachigan, *Statistical Analysis*, Radius Press, New York, 1986.
7. D. Ary and L. Jacobs, *Introduction to Statistics: Purposes and Procedures*, Holt, Reinhart and Winston, New York, 1976.
8. V. Covello and M. Merkhofer, *Risk Assessment Methods: Approaches for Assessing Health and Environmental Risks*, Plenum Press, New York, 1993.
9. R. Sokol and F. Rohlf, *Biometry*, W.H. Freeman and Company, New York, 1995.
10. G. Knoll, *Radiation Detection and Measurement*, John Wiley and Sons, New York, 1979.

CHAPTER 3
Systems of Differential Equations

3.1. A Systems View of the Environment

Chapters 1 and 2 presented a view of the environment as a series of fields distributed across space and evolving in time. While fields provide the ultimate description of the state of the environment, there are times when models can be simplified significantly by ignoring the spatial inhomogeneity of fields throughout a region of space. This simplification is completely valid when the field is homogeneous throughout that region, but it also may be at least partially valid if the effect ultimately of interest (e.g. human health) depends on some average property of the field, such as the mean, rather than on the variability in space.

For example, consider a person living in a field of air concentration such as one of the particle fields discussed in Chapter 1. There might be an interest in predicting the probability of lung cancer for that individual as caused by exposure to these particles. This probability depends in some sense (discussed in later chapters) on the total number of particles inhaled. The rate at which particles are inhaled at any moment depends on the concentration at the specific location in the field occupied by the person at that time. If C(x,y,z) is the concentration of the particles in the air at point (x,y,z), the rate of inhalation of particles at that point, $IR_p(x,y,z)$, is [1]:

$$IR_p(x,y,z) = C(x,y,z) \bullet IR_{air} \tag{3.1}$$

where IR_{air} is the rate of inhalation of the air containing the particles. In this equation, $IR_p(x,y,z)$ would be in units of particles/s if C(x,y,z) were in units of particles/m^3 and IR_{air} were in units of m^3/s.

This person probably does not stay at one point in the field, but rather wanders throughout the space (e.g. moving through rooms of the home, going to work, driving, shopping, etc). The rate at which particles are being inhaled will change as this person moves through the field since C(x,y,z) depends on location. If a large number of individuals move at random through the field, the average number of particles, $N_{inhaled}$, inhaled by individuals in the population during some time interval, Δt, will be:

(3.2) $$N_{inhaled} = \int IR_p(x,y,z)dt$$

where the integration of the right hand side is from the start of the exposure, t, to the end of the exposure, t+Δt.

This integral over time, however, can be replaced by an integral over the three dimensional space of the field of air concentration if the person moves completely at random through the space and spends an equal amount of time in each volume. In this case, the total amount of particles inhaled is equal to the product of the inhalation rate of air and the mean concentration throughout the region of space through which the individual can move. If C_{mean} is this mean concentration, Equation 3.2 is replaced by:

(3.3) $$N_{inhaled} = C_{mean} \bullet IR_{air}$$

Finally, noting that the mean concentration, C_{mean}, volume of the region, V_{air}, and total number of particles in this volume, N_{air}, are related by:

(3.4) $$C_{mean} = N_{air} / V_{air}$$

Equation 3.3 may be written as:

(3.5) $$N_{inhaled} = C_{mean} \bullet IR_{air} = (N_{air} / V_{air}) \bullet IR_{air}$$

Predictions of the total number of inhaled particles under these conditions of random movement, therefore, require only information on the total number of particles contained in the region of space through which an individual moves, rather than the spatial inhomogeneity of the concentration. This is only true, of course, if individuals spend an equal amount of time in each part of the space.

The above example replaces the concept of a field with the concept of a *compartment*. A compartment is a region of space that can be treated as a single entity in a risk assessment because the inhomogeneity of the field property in that region can be replaced by an average or mean. When this is the case, an environment can be viewed as a connected series of compartments, or a *system* of compartments. Rather than focus on the evolution of fields in such cases, the focus is on the evolution of the state of the system. The *state of the system*, in turn, is described by the mean field quantity in each of the compartments comprising the system, or by the total mass or energy in each compartment, at some moment in time.

An example is shown in Figure 3.1, which displays a system encountered commonly in environmental risk assessment. In the example, a pollutant is released from a source into a compartment, and then moves between the other compartments. Each compartment (air, water, etc) is contained in a specific region of space, and each is characterized by a mean concentration of the pollutant at any moment in time, $C_i(t)$, where i is an index referring to the ith compartment. As the pollutant moves between the compartments, $C_i(t)$ evolves in each compartment. Using Equation 3.4, it should also be clear that $N_i(t)$, or the amount of the pollutant in each compartment, evolves.

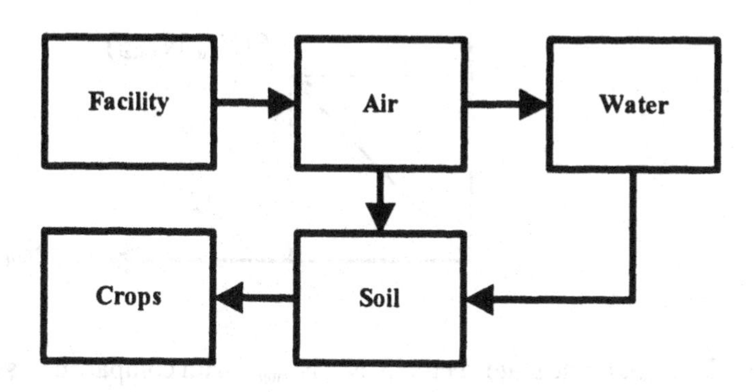

Figure 3.1. An example of a compartmental system in the environment. The boxes represent compartments, each containing a mass of the pollutant. The arrows indicate the direction of flow of the pollutant in the system.

In compartmental models, the state of the environmental system can be represented by a vector; the result is a *state vector*. The number of dimensions of the vector equals the number of compartments. The length of the vector along any dimension equals the amount of the pollutant in the compartment represented by that dimension. For example, $S(N_{air}, N_{water}, N_{crops})$ would be the state of an environmental system consisting of air, water and crops. S(3,7,2) in that case might represent a state of the environmental system in which the air contains 3 kg of the pollutant, the water contains 7 kg, and the crops contain 2 kg. The state of the system could also be expressed using concentrations using Equation 3.4, with the state being given by:

$$(3.6) \qquad S(C_{air}, C_{water}, C_{crops}) = S(N_{air} / V_{air}, N_{water} / V_{water}, N_{crops} / V_{crops})$$

Figure 3.2 shows an example of a state vector for a simple system of 2 compartments. In this example, the state is described completely by a 2-D vector giving the amount of pollutant in each of the two compartments of the system.

A series of compartments becomes a system due to the interconnections between the compartments. Changes in any one compartment result in changes in one or more of the other compartments (usually in all of them). Such systems then are subdivided into two classes [2]:

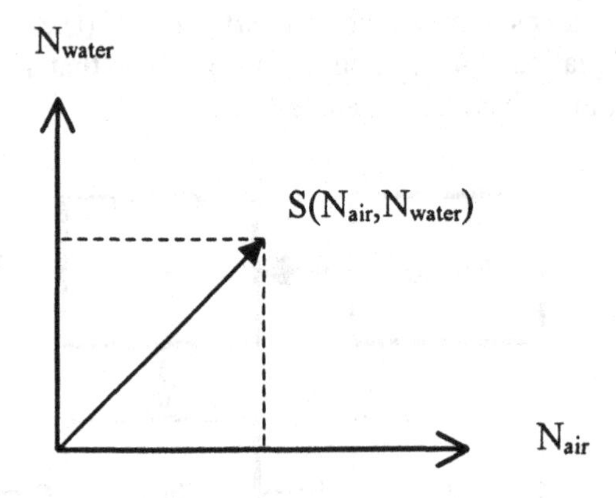

Figure 3.2. An example state vector, $S(N_{air},N_{water})$, with compartments of air and water. The amount of the pollutant in any compartment is given by the projection of the vector onto that axis, as indicated by the vertical and horizontal dashed lines.

- *Closed System*: a series of compartments in which there is no interaction with the world outside the system. The evolution of the state vector is a function only of interactions taking place between the compartments. For example, a pollutant might be produced within a system consisting of lake water, the fish in the lake, and the algae growing in the lake. The pollutant might move from the water to the algae to the fish, but if the system is closed the total amount of pollutant found in the system will not change. Consider the state vector in such a system, with the magnitude of the vector equaling the total amount of the pollutant in the system. The state vector might rotate through the three dimensional space (water, algae and fish), with the components along each axis changing as the evolution takes place, but the magnitude of the vector would remain unchanged.

- *Open System*: a series of compartments in which there is interaction between the compartments of the system and the world outside the system. The evolution of the state vector is a function of both interactions taking place between the compartments and of interactions taking place between the system and the outside world. For example, a pollutant might be introduced (from the outside) into a system consisting of lake water, the fish in the lake, and the algae growing in the lake. Consider the state vector in such a system, with the magnitude of the vector equaling the total amount of the pollutant in the system. The state vector probably will rotate through the three dimensional space (water, algae and fish), with the components along each axis changing as the evolution takes place. In addition, the magnitude of the vector would change with time. If the interactions with the outside world tended to increase the total amount of pollutant in the system, the magnitude of the state vector would increase. If the interactions with the outside world tended to decrease the total amount of pollutant in the system, the magnitude of the state vector would decrease.

Environmental systems evolve due to one or more of several processes that in general control the movement of fate of materials and energy within systems. These processes are the "Three Ts" of environmental systems [3]:

- *Transport*: any process that causes material or energy to move within the space of a given compartment. This does not include movement between compartments. An example of transport is air dispersion, and an example of a transport model is the Gaussian dispersion model of Chapter 1, since the air pollutant moves throughout the atmosphere but still remains in the atmosphere. Transport does not change the total amount of material or energy in a compartment, but rather changes the inhomogeneity of the field.

- *Transfer*: any process that causes material or energy to move from one compartment to another. An example is the movement of a pollutant from the air, to a lake, to fish in a lake, etc. In some sense, the distinction between transfer and transport is artificial, since in both cases the material or energy is moving through space. In addition, the processes of movement are the same (diffusion, sedimentation, etc, as discussed in Chapter 1). The difference between the two is that transfer is an example of transport that moves the material or energy across the boundary between compartments.

- *Transformation*: any process that causes material or energy to change form. An example is chemical transformation or radioactive decay (a form of physical transformation). The transformation takes place at a specific point (x,y,z) in a compartment, and so transformation is not an example of either transport or transfer.

Systems models may describe either transfer or transformation (but are not applicable to transport processes). The mathematical details of transfer and transformation are identical, as will be seen in later sections. As a result, transfer and transformation models often are homologous (identical in mathematical form).

3.2. Mass/Energy Balance and Conservation Laws

At the heart of systems models in environmental risk assessment lies the principle of *conservation of mass and energy*, or *mass/energy balance*. This principle states that, for a closed system, the sum of the energy and mass in the system remains constant, even if both are transported, transferred or transformed. While Einstein demonstrated (through $E=mc^2$) that mass can be converted (transformed) into energy, and energy back into mass, environmental phenomena usually do not involve such conversions on a significant scale. Instead, mass remains mass and energy remains energy, at least within the limits of measurements that are relevant to environmental phenomena. As a result, it is possible in environmental analyses to separate the principle of mass/energy conservation into two principles:

- *Conservation of mass*: the total mass in a closed system remains constant during the evolution of the system. The mass may be transported in the compartments, may transfer between compartments, and may even be transformed, but the total mass in the system remains the same. This is equivalent to saying that the state vector for mass retains a constant magnitude even as it rotates in its space.

- *Conservation of energy*: the total energy in a closed system remains constant during the evolution of the system. The energy may be transported in the compartments, may transfer between compartments, and may even be transformed, but the total energy in the system remains the same. This is equivalent to saying that the state vector for energy retains a constant magnitude even as it rotates in its space.

For an open system, the two principles still apply, but the definitions must accommodate the movement of mass or energy across the boundaries separating the system from the outside world. Instead of mass remaining constant as the open system evolves, the rate of change of mass in the system must equal the rate of *inflow* of mass into the system minus the rate of *outflow* of mass from the system. If N(t) is the total mass in the system at some moment in time, t, then the rate of change of the mass in the system at that moment must be:

$$R(t) = \text{Inflow}(t) - \text{Outflow}(t) \qquad (3.7a)$$

The rate of change of a quantity, $N(t)$, with respect to time is the derivative of $N(t)$ with respect to time. Therefore:

(3.7b) $$R(t) = dN(t)/dt = \text{Inflow}(t) - \text{Outflow}(t)$$

Equations 3.7a and 3.7b also apply to rates of change of energy in an environmental system. The conservation principles may then be generalized as:

- *Conservation of mass*: the rate of change of mass in a system equals the difference between the rate of inflow of mass into the system across the system boundaries and the rate of outflow of mass from the system across the system boundaries. Where this difference is zero, the total mass in the system remains constant in time. Where this difference is positive, the total mass in the system is increasing in time. Where this difference is negative, the total mass in the system is decreasing in time.

- *Conservation of energy*: the rate of change of energy in a system equals the difference between the rate of inflow of energy into the system across the system boundaries and the rate of outflow of energy from the system across the system boundaries. Where this difference is zero, the total energy in the system remains constant in time. Where this difference is positive, the total energy in the system is increasing in time. Where this difference is negative, the total energy in the system is decreasing in time.

Example 3.1. A system consists of the water of a lake, the plants in the lake, and the fish in the lake. A pollutant enters the system at a rate of 3 g/d. The same pollutant leaves the system at a rate of 2 g/day. What is the rate of change of the amount of pollutant in the system, $dN(t)/dt$, at any moment?

Using the conservation of mass Equation 3.7b, and noting that the rate of inflow (3 g/d) and the rate of outflow (2 g/d) are constant in time:

$$dN(t)/dt = \text{Inflow}(t) - \text{Outflow}(t) = 3 - 2 = 1 \text{ g/d}$$

Therefore, the rate of change of mass in the system is positive (the mass is increasing) and equal to 1 g/d.

Equation 3.7b is an example of a *differential equation*. Such equations describe the rate of change of a quantity such as N(t) with respect to some dimension such as time, t. The mathematics of differential equations consists of methods to find a function which has a derivative described by the differential equation. For example, the task in Equation 3.7b involves finding a function, N(t), whose derivative with respect to time satisfies the right hand side of that equation. Ideally, the solution is *unique*; i.e. there is a single function N(t) whose derivative equals the right hand side of the differential equation. Most differential equations encountered in environmental risk assessment have unique solutions; all of the differential equations considered in this chapter have unique solutions.

In practice, the differential equation usually can be satisfied by an entire *class* of functions. For example, consider the following differential equation:

$$dN(t)/dt = -2$$

This might arise from a conservation of mass problem in which the rate of inflow is constant at 5 g/d and the rate of outflow is constant at 7 g/d. Note that N(t) has the property that its slope, or derivative, is negative, so N(t) is decreasing in time at a rate of 2 g/d. There is an infinite number of functions, N(t), with this property, all parallel to each other as shown in Figure 3.3. All of these functions have a property in common (their slope), and have the same mathematical form (a straight line), and so they constitute a *class*. Which of these members of the class is the correct solution to the physical problem being solved (i.e. which equation, N(t), describes the actual system being modeled)?

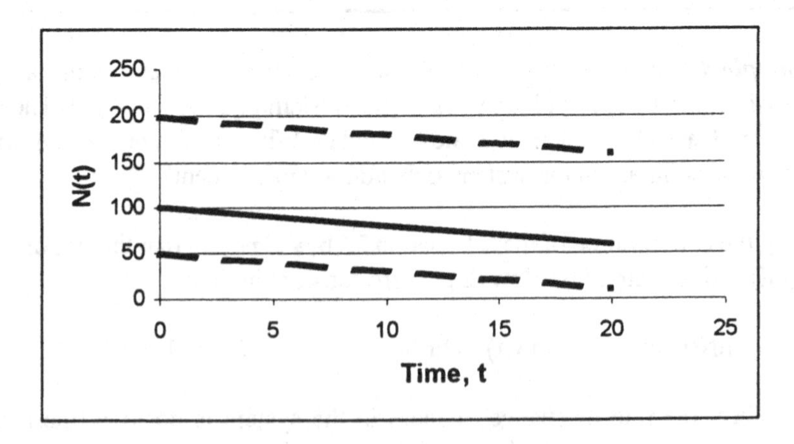

Figure 3.3. The class of functions, N(t), satisfying a differential equation that is constant at -2. Only the middle line solid line, however, satisfies the initial condition that N(0) equals 100.

To narrow the class to the single function describing the physical system, it is necessary to know the value of the function at some moment in time. Typically, this is the value at t equal 0, and so this value is an *initial condition* [4]. In Figure 3.3, the initial condition might be 100 g for the physical system, and so only the solid line (not the dashed lines) would be the unique solution to this problem. Modeling a system described by differential equations such as the ones to appear in this chapter involves establishing the compartmental model with all of its connections of inflow and outflow, writing the defining conservation of mass or energy equations as a series of differential equations, finding the class of functions whose solution satisfies this system of equations, and applying initial conditions to determine the unique solutions.

3.3. Linear Differential Equations

The right hand side of Equation 3.7b contains rates of inflow and outflow. Example 3.1 shows a case in which these rates are constant in time. One might write the defining differential equation as follows:

$$(3.8) \qquad dN(t)/dt = \text{Inflow}(t) - \text{Outflow}(t) = c_1N^0(t) - c_2N^0(t)$$

where c_1 and c_2 are constants equal to 3 and 2, respectively. Each constant is multiplied by $N^0(t)$, which is N(t) raised to the 0th power. Since any number raised to the 0 power equals 1, dN(t)/dt simply equals $c_1 - c_2$. Both terms on the right hand side of Equation 3.8 are examples of *zeroth-order terms* (note the differential equation itself is *first-order* since it involves the first derivative only). The physical (chemical, biological, etc) process to which the term refers then is a *zeroth-order process*. For all such processes, the rate of inflow or outflow (depending on which process is being described) is not a function of N(t).

The idea of the order of a term or process (as distinct from the order of a differential equation) can be seen by generalizing Equation 3.8:

$$(3.9) \qquad dN(t)/dt = \text{Inflow}(t) - \text{Outflow}(t) = c_1N^i(t) - c_2N^i(t)$$

where the rates of inflow and outflow are now functions of N(t) raised to the ith power. In environmental systems, i will be an integer equal to 0, 1, 2, etc. The order of the term or process then is i. For example, a first-order term contains N(t) raised to the first power, so the rate of inflow or outflow into a compartment at a moment in time is always proportional to the amount of mass or energy in the compartment at that moment in time. This chapter considers mixtures of zeroth and first-order terms, all appearing within first-order differential equations. Such differential equations also are linear; non-linear differential equations involve higher-order terms on the right-hand side of Equation 3.9.

First-order differential equations contain terms on the right hand side equal to the product of a constant and N(t). The constant is a *proportionality constant*. More specifically, it is a *first-order rate constant*. In the case of transfer between compartments, the constant is a *first-order transfer rate constant*. Such rate constants typically are given by the symbol k or λ. In this book, a first-order transfer rate constant will be shown as λ. Since the same differential equations may be used to describe transformation processes (also with zeroth-order, first-order, etc terms), there is an analogous set of *first-order transformation rate constants*. These will be shown in the book as k.

The units of first-order rate constants (λ or k) are inverse time (e.g. s^{-1} or d^{-1}). The numerical value of the rate constant equals the fraction of the mass or energy being transferred or transformed per unit time. For example, a first-order rate constant, λ, of 0.3 d^{-1} means 30% of the mass or energy in a compartment is being transferred per day. This is an instantaneous transfer rate constant; it does not mean literally that 30% of the mass in a compartment at the start of a day will be transferred during that day, since as this transfer occurs N(t) will change and so the rate of transfer (in g/d) also will change.

Example 3.2. A compartment contains 6 ergs of energy at some moment in time, t, equal to 5 minutes. There is no rate of inflow, and the rate of outflow is described by a first-order transfer process. The first-order transfer rate constant, λ, equals 0.1 per minute. The differential equation for this problem is:

$$dN(t)/dt = -\lambda N(t) = -0.1N(t)$$

The rate at which energy is leaving the compartment at t equal 5 minutes then is:

$$\text{Outflow}(5) = 0.1N(5) = 0.1 \bullet 6 = 0.6 \text{ ergs/minute}$$

Since this chapter considers systems of compartments with transfer between the compartments, or systems of different forms of material or energy with transformation between them, it is necessary to specify the transfer or transformation processes more fully in giving the transfer or transformation rate constants. In general, we will show a transfer rate constant as λ_{ij}, where i is the subscript indicating the compartment *from which* the material or energy is flowing and j is the subscript indicating the compartment *to which* material or energy is flowing. R_{ij} is the rate at which the material or energy transfers from compartment i to compartment j. Since transfer is from the ith to the jth compartment, the defining equation for such a first-order process is:

$$(3.10) \qquad R_{ij}(t) = \lambda_{ij} N_i(t)$$

or the rate of transfer from compartment i to compartment j at some moment in time equals the product of the transfer rate constant from i to j and the amount of material or energy in compartment i at that time.

For example, $R_{1,5}$ indicates the rate at which material or energy is transferring from compartment 1 to compartment 5 and $\lambda_{1,5}$ is the rate constant for the transfer between these compartments. As a more specific example, consider the rate of transfer from the air to the water in Figure 3.1, $R_{air,,water}$. If that process is first-order, the transfer rate constant would be shown as $\lambda_{air,water}$ and would equal the fraction of the material or energy in the air that is moving to the water per unit time. Using Equation 3.10, $R_{air,water}(t)$ equals $\lambda_{air,water}$ times $N_{air}(t)$. Similarly, if there is a chain of chemicals produced by sequential transformations (from chemical 1 to chemical 2 to chemical 3, etc), the transformation rate constant for transformations from the ith to the jth chemical would be k_{ij}.

3.4. Systems of Differential Equations

The tools are now in place to analyze the evolution of the state vector for systems of compartments such as the one shown in Figure 3.1. In that system, there is transfer from a facility to the air, from the air to the water and to the soil, from the water to the soil, and from the soil to the crops. The general form of the differential equation for this system using Equation 3.7b is:

$$(3.11) \qquad dN_i(t)/dt = \text{Inflow}(t) - \text{Outflow}(t) = \Sigma R_{ji}(t) - \Sigma R_{ij}(t)$$

where $N_i(t)$ is the amount of material or energy in compartment i at time t; $R_{ji}(t)$ is the rate of transfer from compartment j to compartment i at time t; and $R_{ij}(t)$ is the rate of transfer from compartment i to compartment j at time t. The summation associated with $R_{ji}(t)$ is over all compartments transferring material or energy to compartment i, while the summation associated with $R_{ij}(t)$ is over all compartments to which compartment i is transferring material or energy.

Equation 3.11 describes a closed system. If the system is open, there may be a rate of movement from outside the system to one or more of the compartments inside the system, as well as movement back out of the system from one or more of the compartments. In general, we will show such rates as $R_{in,i}(t)$ for rates into compartment i from outside the system, and $R_{i,out}(t)$ for rates out of the system through compartment i. Equation 3.11 then becomes:

$$(3.12) \qquad dN_i(t)/dt = [R_{in,i}(t) + \Sigma R_{ji}(t)] - [R_{i,out}(t) + \Sigma R_{ij}(t)]$$

Example 3.3. Consider the system of compartments in Figure 3.1. Using Equation 3.11, the five differential equations describing this system are:

$$dN_{facility}(t)/dt = -R_{facility,air}(t)$$

$$dN_{air}(t)/dt = R_{facility,air}(t) - [R_{air,water}(t) + R_{air,soil}(t)]$$

$$dN_{water}(t)/dt = R_{air,water}(t) - R_{water,soil}(t)$$

$$dN_{soil}(t)/dt = [R_{air,soil}(t) + R_{water,soil}(t)] - R_{soil,crops}(t)$$

$$dN_{crops}(t)/dt = R_{soil,crops}(t)$$

Note that the system can also be treated as a closed system by identifying where the material or energy is coming from or going to outside the system, and including this location as a compartment within the system. Almost any open system can be reduced in this manner to a closed system, unless the universe as a whole ultimately is open.

Example 3.4. In Example 3.3, add a rate into the crops from outside the system. The new differential equation for that compartment is:

$$dN_{crops}(t)/dt = R_{in,crops}(t) + R_{soil,crops}(t)$$

The other differential equations for the system remain the same as in Example 3.3.

Some of the processes in Equation 3.12 may be zeroth-order and some may be first-order (or even higher orders, although these are not considered here). The appropriate forms of the various rate terms will depend on the order of the processes. If they all are zeroth-order, the functions describing the rates (the rates may be a function of time even if they are not functions of the amount in the compartments) are placed into Equation 3.12. If all processes are first-order, the rates are proportional to the amounts in the compartments at any moment in time and Equation 3.11 becomes (for a closed system):

$$dN_i(t)/dt = \Sigma R_{ji}(t) - \Sigma R_{ij}(t) = \Sigma \lambda_{ji} N_j(t) - \Sigma \lambda_{ij} N_i(t) \quad (3.13)$$

Example 3.5. Revisiting Example 3.3, assume all rates of transfer are first-order. Using Equation 3.13, the five differential equations describing this system are:

$$dN_{facility}(t)/dt = -\lambda_{facility,air} N_{facility}(t)$$

$$dN_{air}(t)/dt = \lambda_{facility,air} N_{facility}(t) - [\lambda_{air,water} N_{air}(t) + \lambda_{air,soil} N_{air}(t)]$$

$$dN_{water}(t)/dt = \lambda_{air,water} N_{air}(t) - \lambda_{water,soil} N_{water}(t)$$

$$dN_{soil}(t)/dt = [\lambda_{air,soil} N_{air}(t) + \lambda_{water,soil} N_{water}(t)] - \lambda_{aoil,crops} N_{soil}(t)$$

$$dN_{crops}(t)/dt = \lambda_{aoil,crops} N_{soil}(t)$$

There are two kinds of systems of differential equations encountered in risk assessment. *Uncoupled systems*, or *catenary systems*, involve transfer or transformation in one direction only. Figure 3.1 displays such a catenary system. Once material or energy transfers or transforms from one compartment to another, there is no flow backwards to the first compartment. The flow is down through the system of compartments, from the first to the last compartment in the chain. By contrast, *coupled systems* involve flow in both directions. The method of solution for systems of differential equations used in this chapter applies only to uncoupled, or catenary, systems (although they may be open or closed). Coupled systems are considered in Chapters 5 and 6. Most real systems are coupled to at least some extent, although many may be approximated by uncoupled systems since flow is predominantly in one direction.

Many methods exist for solving uncoupled systems of differential equations with zeroth and first-order terms. Rather than considering all of them, we will focus on one method, *Bernoulli's method* [4], which will work in all cases considered here. The method may take longer to apply in some cases than more specialized methods, but it has the merit of applying to any set of uncoupled differential equations with zeroth and/or first-order terms (it also applies to higher-order terms, but these are not considered here). As a result, it is applicable to a wide range of systems encountered in environmental risk assessment.

Bernoulli considered the class of differential equations describing an arbitrary function F(x) given by:

(3.14) $$dF(x)/dx = Q(x) - P(x)F(x)$$

where x may be any variable (such as time, t) and P(x) and Q(x) are any functions of x (including zeroth-order or first-order functions). Notice that Equations 3.11 through 3.13 all are special cases of this general form of Bernoulli's equation (Equation 3.14). If a differential equation can be placed into this form, the solution is:

(3.15) $$F(x) \bullet e^{a} = \int Q(x)\, e^{a}\, dx + F(0)$$

where the integral on the right hand side is from 0 to x and where:

(3.16) $$a = \int P(x)\, dx$$

and again, the integral is from 0 to x. F(0) is the initial value for this function (or the value of the function at x equal 0).

Note: To be more correct mathematically, the x in Equation 3.16 should be replaced by another symbol such as x*. The integral in that equation then is still from 0 to x, but the equation would be shown as:

$$a = \int P(x^*)\, dx^*$$

This convention will not be employed in the following discussion. The reader should, however, keep this issue in mind.

3.5. Applications of Bernoulli's Method

This section reviews a series of increasingly more complex problems of predicting systems evolution appearing in problems of environmental risk assessment. The examples all use Bernoulli's method, and a large class of environmental systems models can be solved by combinations of the following examples.

3.5.1. Constant rates of inflow and outflow

By far the simplest phenomenon is one in which both the inflow and outflow rates are constant in time for a compartment. Let c_1 be the inflow rate and c_2 be the outflow rate for that compartment. The differential equation for this compartment then is:

$$dN(t)/dt = c_1 - c_2 \tag{3.17}$$

Bernoulli's method made be used by equating x with t, equating F(x) with N(t), and equating Q(x) with $(c_1 - c_2)$ in Equation 3.14. P(x) in Equation 3.14 is 0. With P(x) = 0, a in Equation 3.14 also is 0 (see Equation 3.16). Equation 3.15 then yields:

$$N(t) \bullet e^0 = \int (c_1 - c_2)\, e^0\, dt + N(0) \tag{3.18}$$

Since e^0 is 1, we find:

$$N(t) = \int (c_1 - c_2)\, dt + N(0) \tag{3.19}$$

However, $(c_1 - c_2)$ is a constant and can be brought outside the integral, yielding:

$$N(t) = (c_1 - c_2) \int dt + N(0) \tag{3.20}$$

This integral is simply t evaluated from 0 to t, or (t-0), or t. Therefore, Equation 3.20 becomes:

$$N(t) = (c_1 - c_2) \bullet t + N(0) \tag{3.21}$$

Examples of this equation are shown in Figure 3.4.

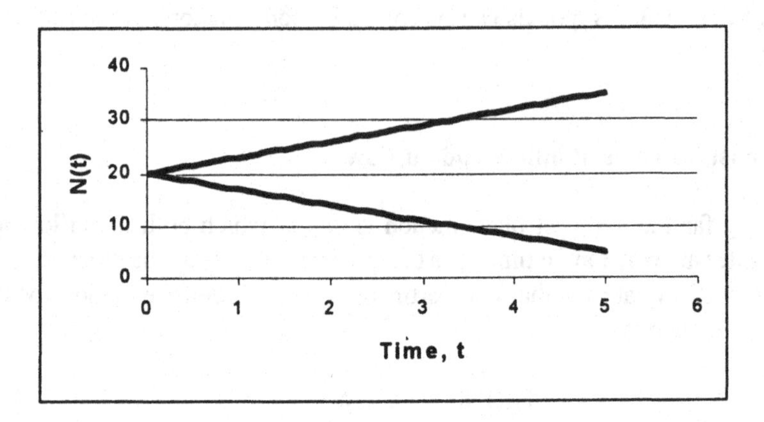

Figure 3.4. Two examples of Equation 3.21. In both examples, N(0) is 20. In the increasing curve (positive slope), $(c_1 - c_2)$ is 3. In the decreasing curve (negative slope), $(c_1 - c_2)$ is –3.

3.5.2. Exponential washout

Consider a case in which some material or energy is placed into a compartment at t equal 0. This initial amount is N(0). There is no further inflow into the compartment over time. The material or energy is removed from the compartment by a first-order process with transfer rate constant λ. The differential equation for this compartment then is:

(3.22) $$dN(t)/dt = 0 - \lambda N(t)$$

Bernoulli's method made be used by equating x with t, equating F(x) with N(t), setting Q(x) to 0, and equating P(x) with λ in Equation 3.14. Using Equation 3.16, a is found from:

(3.23) $$a = \int P(t)\,dt = \int \lambda\,dt = \lambda t - \lambda 0 = \lambda t$$

Equation 3.15 then yields:

(3.24) $$N(t) \bullet e^{\lambda t} = \int 0\, e^{\lambda t}\,dt + N(0) = 0 + N(0)$$

Example 3.6. Consider a compartment with constant rates of inflow and outflow equal to 2 g/d and 5 g/d, respectively. The amount of mass in the compartment at t equal 0 is 20 g. How much mass is present at 3 days after the start of the system?

Using Equation 3.21:

$$N(t) = (c_1 - c_2) \bullet t + N(0) = (2 - 5) \bullet t + 20 = -3t + 20$$

At t equal 3 days, we find:

$$N(3) = -3 \bullet 3 + 20 = -9 + 20 = 11 \text{ g}$$

Dividing both sides by $e^{\lambda t}$ yields:

$$N(t) = N(0) / e^{\lambda t} = N(0)\, e^{-\lambda t} \tag{3.25}$$

An example of this solution may be seen in Figure 3.5. The solution is referred to as *exponential washout* because the initial amount in the compartment decreases exponentially with time. Note that N(t) never becomes exactly 0, instead approaching 0 asymptotically.

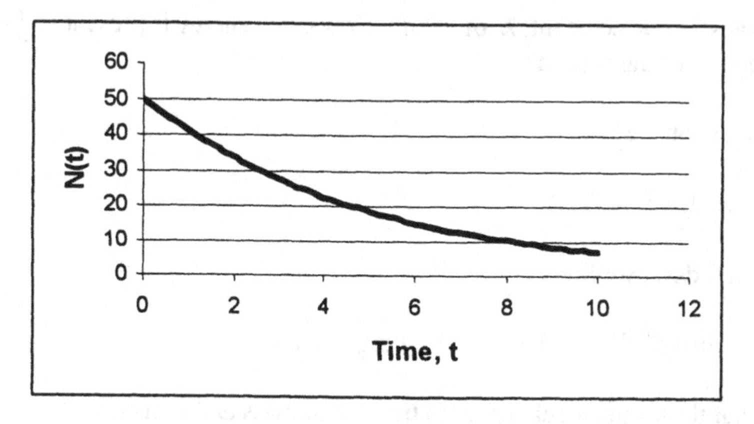

Figure 3.5. An example of exponential washout described by Equation 3.25. In this example, N(0) is 50 and the transfer rate constant, λ is 0.2 d^{-1}.

Figure 3.5 and Equation 3.25 may be used to illustrate the concept of a removal *half-life* [5]. We might ask how long it takes for half of the material placed into a compartment at t equal 0 to be removed. This will be true when N(t) in Equation 3.25 equals half the value of N(0). Solving that equation under these conditions yields:

(3.26a) $$N(t) = N(0)\, e^{-\lambda t} = 0.5N(0)$$

or

(3.26b) $$N(t) / N(0) = e^{-\lambda t} = 0.5$$

Taking the natural logarithm of all terms, and noting that $\ln(e^a) = a$:

(3.27) $$\ln(e^{-\lambda t}) = -\lambda t = \ln(0.5) = -0.693$$

or

(3.28) $$t = T_{1/2} = 0.693 / \lambda$$

Example 3.7. Consider a compartment with an initial amount of energy equal to 50 ergs. The energy is transferred out of the compartment by a first-order process with a transfer rate constant, λ, of 0.2 d^{-1}. How much energy is present at 5 days after the start of the system?

Using Equation 3.25:

$$N(t) = N(0)\, e^{-\lambda t} = 50\, e^{-0.2t}$$

At t equal 5 days, we find:

$$N(5) = N(0)\, e^{-0.2 \bullet 5} = 50\, e^{-1} = 18.4 \text{ ergs}$$

Notice that the same models apply to transfer of mass and energy.

In this relationship (Equation 3.28), $T_{1/2}$ is the half-life, or the value of t satisfying the requirement that N(t) equal 50% of N(0) in a case of simple exponential washout. In many cases, values of the half-life, rather than of λ, are reported for environmental compartments and processes. For Figure 3.5, the half-life appears visually to be between 3 and 4 days. Using Equation 3.28, it is found to be 3.5 days.

3.5.3. Constant inflow, exponential washout

Consider a case in which some material or energy is transferred into a compartment at a constant rate, R_{in}. The material or energy is removed from the compartment by a first order process with transfer rate constant λ. At t equal to 0, the initial value is N(0). The differential equation for this compartment is:

(3.29) $$dN(t)/dt = R_{in} - \lambda N(t)$$

Bernoulli's method made be used by equating x with t, equating F(x) with N(t), equating Q(x) with R_{in}, and equating P(x) with λ in Equation 3.14. Using Equation 3.16. a is found from:

(3.30) $$a = \int P(t)\, dt = \int \lambda\, dt = \lambda t$$

Use of equation 3.15 then yields:

(3.31) $$N(t) \bullet e^{\lambda t} = \int R_{in}\, e^{\lambda t}\, dt + N(0) = R_{in} \int e^{\lambda t}\, dt + N(0)$$

The integral is evaluated from 0 to t and may be found from the general relationship (which will be used repeatedly in this Chapter) [6]:

(3.32) $$\int e^{\lambda t}\, dt = (e^{\lambda t} - e^{\lambda 0}) / \lambda$$

for cases where the integral is from 0 to t, as is the case here. Substituting Equation 3.32 into Equation 3.31 yields:

(3.33) $$N(t) \bullet e^{\lambda t} = (R_{in}\, (e^{\lambda t} - e^{\lambda 0}) / \lambda) + N(0)$$

Dividing both sides by $e^{\lambda t}$ yields:

(3.34) $$N(t) = (R_{in}\, (1 - e^{-\lambda t}) / \lambda) + N(0)\, e^{-\lambda t}$$

In obtaining Equation 3.34, the following relationship was used:

$$(3.35) \quad (e^{\lambda t} - e^{\lambda 0})/e^{\lambda t} = (e^{(\lambda t - \lambda t)} - e^{(\lambda 0 - \lambda t)}) = (e^{(0)} - e^{(-\lambda t)}) = (1 - e^{-\lambda t})$$

An example of this solution may be seen in Figure 3.6. Note that N(t) approaches the value R_{in}/λ asymptotically.

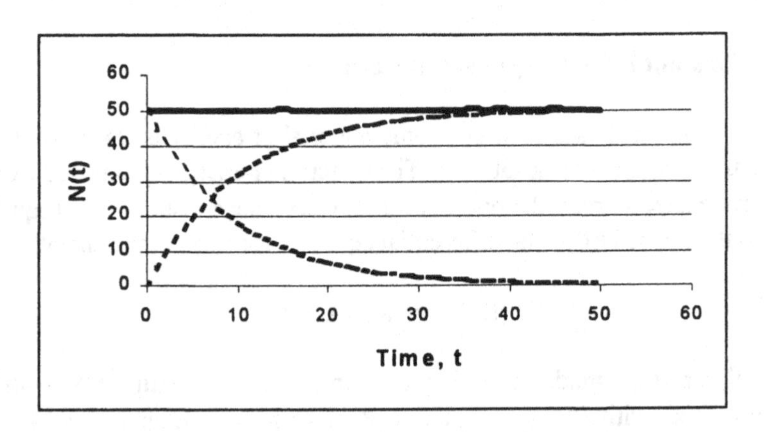

Figure 3.6. An example of a constant rate into a compartment with a first-order process out, using Equation 3.34. The rising dashed line is the first term on the right hand side (in brackets) representing the retention of material flowing into the compartment. It is the value N(t) would have if N(0) were 0. The falling dashed line is the second term representing exponential washout of the initial material. It is the value N(t) would have if R_{in} were 0. The solid horizontal line is the total mass in the compartment, equal to the sum of the two dashed lines. Note the asymptotic approach of the rising dashed line to 50, which equals R_{in}/λ.

Equation 3.34 and Figure 3.6 illustrate the principle of *steady state*. When a system reaches steady state, the value of $N_i(t)$ remains constant in time for all compartments in the system. In other words, the state vector becomes constant or steady; hence the term steady state. *Equilibrium* in a compartment is a special case of steady state. Returning to Equation 3.7b, it may be noted that the derivative of N(t) is 0 when the inflow rate equals the outflow rate. When this balance occurs, the compartment is said to be in equilibrium. Most examples of steady state also are examples of equilibrium, although it is possible to have steady state without a balancing of the rates of transfer into and out of a compartment. Any difference between inflow and outflow might be balanced by a transformation process internal to the compartment. In that case, the system would still be in steady state but not in equilibrium. Steady state might apply in a system even if equilibrium between inflow and outflow by transfer is not established.

The physical interpretation of equilibrium may be seen from the rising dashed line in Figure 3.6 and the differential equation describing it (Equation 3.29). R_{in} for that equation is constant in time. The product of λ and N(t), however, begins at 0 if N(0) is 0 and rises as N(t) increases due to inflow. Eventually, N(t) gets sufficiently large that the product of λ and N(t) equals R_{in}. At that time, the derivative of N(t) becomes 0 and the compartment has reached equilibrium.

Full equilibrium never is reached in an environmental system. This may be seen by noting that the rising dashed line in Figure 3.6 never reaches 50 but approaches asymptotically. Still, a system might be "close enough" to equilibrium to be considered at equilibrium. There is no length of time generally considered to yield an "effective equilibrium", but we can construct a general idea of this time by considering Equation 3.34. Note that the term involving N(0) goes to 0 at large values of t. Considering only the other term, we might ask how long it will take for this term to approach to within a fraction X of its asymptotic value of R_{in}/λ. In other words, we want to solve the equation:

(3.36a) $$(R_{in}(1 - e^{-\lambda t}) / \lambda) = X \bullet R_{in}/\lambda$$

or

(3.36b) $$(1 - e^{-\lambda t}) = X$$

or

(3.36c) $$e^{-\lambda t} = 1 - X$$

Taking the natural logarithm of both sides, and noting that $\ln(e^a) = a$, we find:

(3.36d) $$-\lambda t = \ln(1-X)$$

or

(3.36e) $$t = -\ln(1-X) / \lambda$$

It is useful to re-formulate this as a function using half-lives (see Equation 3.28):

(3.36e) $$t = -T_{1/2} \bullet \ln(1-X) / 0.693$$

or

(3.36f) $$t / T_{1/2} = -\ln(1-X) / 0.693$$

The ratio of $t/T_{1/2}$ equals the number of half-lives that have passed by time t.

A graph of this Equation 3.36f is shown in Figure 3.7. From this figure, it is possible to determine the number of half-lives that must pass to be within a fraction X of equilibrium. Note that X approaches 1 for values of t above 6 or 7 half-lives. An effective equilibrium often is reached, therefore, within something on the order of 7 half-lives.

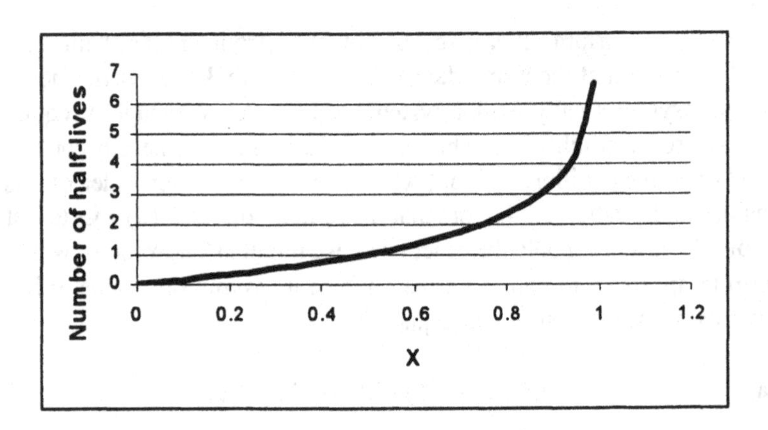

Figure 3.7. Equation 3.36f showing the number of half-lives (y-axis) needed for N(t) to approach within a factor of X of equilibrium for a compartment with constant inflow and a first-order removal process. Note the steepness of the curve above 6 or 7 half-lives.

3.5.4. Exponential washout with multiple pathways

Sections 3.5.1 through 3.5.3 consider cases in which there is a single removal pathway from a compartment. What of the case where there are multiple pathways from the same compartment, with each pathway characterized by a first-order rate constant and with independence between the pathways? In such cases, there will be competition between the pathways for removal of each unit of mass or energy in the compartment.

The solution to this case follows along the lines developed in Section 3.5.2. Consider some material or energy placed into a compartment at t equal 0. This initial amount is N(0). There is no further inflow into the compartment over time. The material or energy is removed from the compartment by a total of m first-order processes, each with transfer rate constant λ_n (the n is a subscript for the nth removal process). The differential equation for this compartment then is:

$$(3.37) \qquad dN(t)/dt = 0 - \lambda_1 N(t) - \lambda_2 N(t) - \ldots \lambda_m N(t)$$

$$= 0 - \Sigma \lambda_n N(t)$$

where the sum is over all values of n from 1 to m.

Example 3.8. Consider a compartment with an initial amount of Chemical A equal to 50 g. Chemical A is transformed to Chemical B by a first-order process with a transformation rate constant, k, of 0.1 hr^{-1}. Chemical A also enters the compartment at a constant rate, R_{in}, of 5 g/hr. How much of Chemical A is present in the compartment at 12 hours after the start of the system?

Using Equation 3.34 (with λ replaced by k since this is transformation):

$$N(t) = (R_{in} (1 - e^{-kt}) / k) + N(0) e^{-kt}$$

$$= (5 (1 - e^{-0.1t}) / 0.1) + 50 e^{-0.1t}$$

At t equal 12 hours, we find:

$$N(12) = (5 (1 - e^{-0.1 \bullet 12}) / 0.1) + 50 e^{-0.1 \bullet 12} = 34.9 + 15.1 = 50 \text{ g}$$

Notice that the same models applied to transfer of mass and energy may be applied to transformation. Note also that, in this example, the amount present at any time equals 50 g. In this particular example, the amount remaining from past inflow exactly balances the loss of mass from the initial value. This is not in general the case. It arises here since N(0) exactly equals R_{in}/k.

Bernoulli's method made then be used in an identical manner as in Section 3.5.2 by equating x with t, equating F(x) with N(t), equating Q(x) with 0, and equating P(x) with $\Sigma\lambda_n$ in Equation 3.14. Using Equation 3.16. a is found from:

$$(3.38) \qquad a = \int P(t)\, dt = \int \Sigma \lambda_n\, dt = \Sigma \lambda_n t$$

Equation 3.15 then yields:

(3.39) $$N(t) \bullet e^{\Sigma\lambda_r t} = \int 0 \, e^{\Sigma\lambda_r t} dt + N(0) = 0 + N(0)$$

Dividing both sides by $e^{\Sigma\lambda_r t}$ yields:

(3.40) $$N(t) = N(0) / e^{\Sigma\lambda_r t} = N(0) \, e^{-\Sigma\lambda_r t}$$

Note that this is the same solution as Equation 3.25, with the single rate constant of Equation 3.25 replaced by the sum of the rate constants.

3.5.5. Subcompartments

Figure 3.5 shows exponential washout for a single compartment, plotted on linear axes. If an exponential washout function is plotted with the y–axis in logarithmic units, Figure 3.8 is obtained. Note that the function, N(t), in this case appears linear (since the natural logarithm of an exponential function is linear).

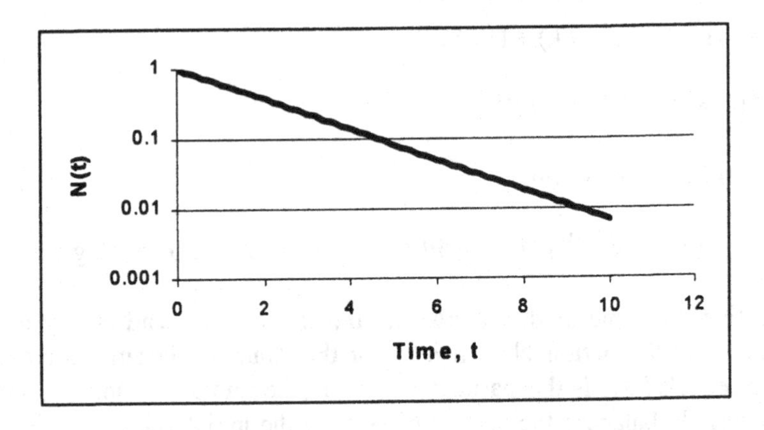

Figure 3.8. A plot of exponential washout, defined by Equation 3.25 or 3.40, with the y-axis logarithmic and the x-axis linear. Note the resulting curve decreases linearly with time on these axes. The rate constant, λ, is 0.5. This figure is normalized to N(0), so the y-axis shows the *fraction* of the initial remaining material in the compartment as a function of time.

Compare Figure 3.8 with the graph in Figure 3.9. One major difference is evident visually. While the graph in Figure 3.8 is a single straight line, the graph in Figure 3.9 appears to contain two distinct regions. The first (at small values of

t) decreases rapidly with t. The second (at large values of t) decreases less rapidly with t. How is this to be explained and modeled?

There are several possible explanations. One is that the removal from the compartment simply cannot be described by first-order processes. The process might be non-linear, with the rate constant, λ, being a function of time (a possibility considered in Section 3.5.7). Another possibility is that what appears to be a single compartment might, in fact, be two compartments, each with a different first-order process. Since these two compartments cannot be distinguished in measurements, we will refer to them here as *subcompartments*. Figure 3.9 then must present predictions of the total amount of material or energy in the two subcompartments; it must depict the sum of the amounts in these subcompartments.

Suppose this explanation is correct. Imagine a total amount of material or energy equal to N(0) being placed in the compartment at time t equal to 0. This material or energy is divided immediately between the two subcompartments, with a fraction f_1 going to the first subcompartment and a fraction f_2 going to the second subcompartment. As a result, the amount of material or energy going initially into the first subcompartment is $f_1N(0)$, and the amount of material or energy going initially into the second subcompartment is $f_2N(0)$.

Once in the first subcompartment, the material or energy is removed by a first-order process with rate constant λ_1. The equation describing the amount of material or energy in this subcompartment as a function of time is the same as Equation 3.25, with N(t) replaced by $N_1(t)$, N(0) replaced by $f_1N(0)$, and λ replaced by λ_1:

(3.41) $$N_1(t) = N_1(0) / e^{\lambda_1 t} = f_1 N(0)\, e^{-\lambda_1 t}$$

Similarly for the second subcompartment:

(3.42) $$N_2(t) = N_2(0) / e^{\lambda_2 t} = f_2 N(0)\, e^{-\lambda_2 t}$$

Since the total amount of material or energy in the two subcompartments must be determined, N(t) will be governed by the following equation:

(3.43) $$N(t) = N_1(t) + N_2(t) = f_1 N(0)\, e^{-\lambda_1 t} + f_2 N(0)\, e^{-\lambda_2 t}$$

Generalizing Equation 3.43 to the case of n subcompartments:

(3.44) $$N(t) = \Sigma N_i(t) = \Sigma f_i N(0)\, e^{-\lambda_i t}$$

where the sums are over all n subcompartments (i.e, from i equal 1 to n).

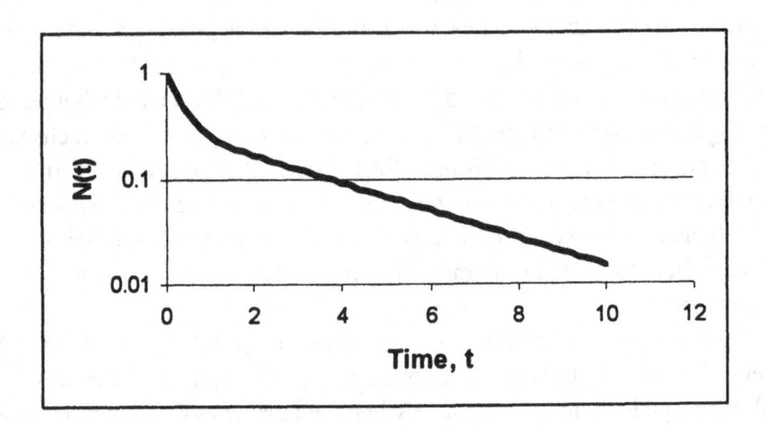

Figure 3.9. A plot of exponential washout from a compartment with two subcompartments, defined by Equation 3.43, with the y-axis logarithmic and the x-axis linear. Note the resulting curve decreases linearly with time for values of t between 0 and 1, and then linearly with a more shallow slope for values of t above 2. There are two compartments in this model. The first compartment contains 70% of the initial material, which is removed from the subcompartment with a rate constant of 3. The second subcompartment contains 30% of the initial material, which is removed from the subcompartment with a rate constant of 0.3. This figure is normalized to N(0), so the y-axis shows the fraction of the initial material in the compartment.

The numerical values for f_i and λ_i may be obtained from different points on the graph in Figure 3.9. Note that f_2 is the fraction of material or energy in the second subcompartment initially (at time, t, equal 0). We will let the second subcompartment be identified as the one with the longest half-life or smallest value of λ. This subcompartment, therefore, is predicted by the part of the curve in Figure 3.9 at large values of t. It appears that this subcompartment dominates N(t) at values of t greater than 2, so let's consider only the curve beyond t equal 2. Fitting a straight-edged ruler to that part of the curve, extrapolate back to t equal 0. This line will cross the y-axis at f_2, which in this case equals 0.3. The value of f_1 then must equal $1-f_2 = 1 - 0.3 = 0.7$.

The numerical values of the rate constants may be obtained from the parts of the curve for N(t) where each subcompartment dominates. As an example, consider the value of λ_2. Again, we will use the part of the curve beyond t equal to

2. Consider the points at t equal to 3, where N(3) is 0.12, and at t equal to 6, where N(6) is 0.05. Between these two points, N(t) will decrease exponentially with the rate constant λ_2. During this decrease, the defining equation will be Equation 3.25, with N(t) replaced by N(6), N(0) replaced by N(3), λ replaced by λ_2, and t replaced by 6-3 or 3.

(3.45) $$N(6) = N(3) / e^{\lambda_2(6-3)} = N(3)\, e^{-\lambda_2 3}$$

which may be re-arranged, using methods from previous sections, to yield:

(3.46) $$\lambda_2 = -\ln(N(6)/N(3)) / 3 = = -\ln(0.05/0.12) / 3 = 0.3$$

The same method may be used to determine λ_1 using values of t below 1. In general, the rate constant associated with a particular region of a curve such as that in Figure 3.9 may be found from:

(3.47) $$\lambda_i = -\ln(N(t_2)/N(t_1)) / (t_2-t_1)$$

where t_1 and t_2 are two times in that region of the curve, t_2 is greater than t_1, $N(t_1)$ is the amount of material or energy in the compartment at time t_1, and $N(t_2)$ is the amount of material or energy in the compartment at time t_2.

This method for finding the rate constants suffers from making use of only several values of N(t), which means many data points may be ignored. If it is desired to use all of the data points, a better approach is to find the best-fitting values of the rate constants using a procedure such as the least squares or maximum likelihood procedures discussed in Chapter 2. Equation 3.44 would be used, with the model predictions compared against the measured values of N(t) at those same times. Different values of the rate constants then would be tried sequentially until the est goodness-of-fit measure was obtained. This approach can, however, be problematic if there are more than two subcompartments, since there may be too many parameters to estimate reliably.

3.5.6. Series of compartments

We return now to the original problem posed in this chapter: how may we model systems of compartments with transfer between them? The building blocks were presented in Sections 3.5.1 through 3.5.5. We apply these here to an example problem involving a catenary system of three compartments with inflow into the first compartment from outside the system, and outflow from the third compartment to outside the system. The system is shown in Figure 3.10.

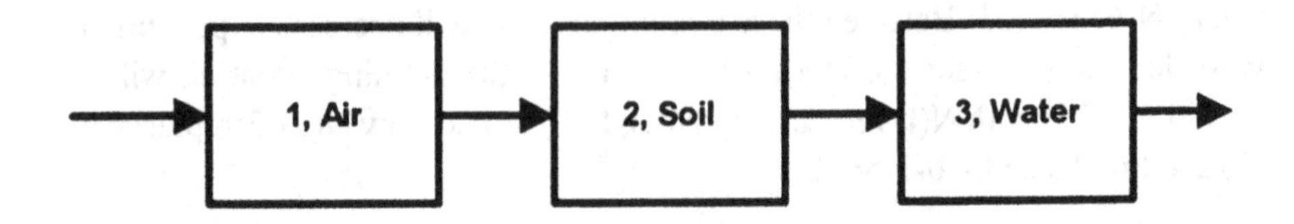

Figure 3.10. The example catenary compartmental system used in Section 3.5.6, arising perhaps from particles entering the air, settling onto soil, and then being washed into lake water by rain. Material moves from outside the system into Compartment 1 (air) at a constant rate, R_{in}. Material moves from Compartment 3 (water) to some location outside the system by a first-order transfer process. Both the transfers from air to soil and from soil to water are first-order.

For the system in Figure 3.10, we will refer to the compartments by their numbers (1 for air, 2 for soil and 3 for water). The rate into Compartment 1 is constant at the value R_{in}. All other arrows in Figure 3.10 represent first-order transfer processes. The first-order transfer rate constant from air to soil is $\lambda_{1,2}$, the first-order transfer rate constant from soil to water is $\lambda_{2,3}$, and the first-order transfer rate constant from water to outside the system is $\lambda_{3,out}$. The defining differential equations for the three compartments then are:

$$\text{(3.48)} \qquad dN_1(t)/dt = R_{in} - \lambda_{1,2}N_1(t)$$

$$\text{(3.49)} \qquad dN_2(t)/dt = \lambda_{1,2}N_1(t) - \lambda_{2,3}N_2(t)$$

$$\text{(3.50)} \qquad dN_3(t)/dt = \lambda_{2,3}N_2(t) - \lambda_{3,out}N_3(t)$$

The initial values in the three compartments are $N_1(0)$, $N_2(0)$ and $N_3(0)$.

The solution to Equation 3.48 was generated in Section 3.5.3. Equation 3.34 is modified here to reflect the terminology of this example:

$$\text{(3.51)} \qquad N_1(t) = (R_{in}(1 - e^{-\lambda_{1,2}t}) / \lambda_{1,2}) + N_1(0)\, e^{-\lambda_{1,2}t}$$

Substituting Equation 3.51 into Equation 3.49 yields:

$$\text{(3.52)} \qquad dN_2(t)/dt = \lambda_{1,2}[(R_{in}(1 - e^{-\lambda_{1,2}t}) / \lambda_{1,2}) + N_1(0)\, e^{-\lambda_{1,2}t}] - \lambda_{2,3}N_2(t)$$

which may be solved with Equation 3.15 as follows:

(3.53) $N_2(t) \bullet e^a = \int \lambda_{1,2}[(R_{in}(1 - e^{-\lambda_{1,2}t}) / \lambda_{1,2}) + N_1(0)\, e^{-\lambda_{1,2}t}]\, e^a\, dt + N_2(0)$

where $a = \int \lambda_{2,3} dt = \lambda_{2,3}t$ using methods introduced in earlier sections and where the integral is from 0 to t. Therefore, Equation 3.53 may be written as:

(3.54) $N_2(t)e^{\lambda_{2,3}t} = \int \lambda_{1,2}[(R_{in}(1 - e^{-\lambda_{1,2}t}) / \lambda_{1,2}) + N_1(0)\, e^{-\lambda_{1,2}t}]\, e^{\lambda_{2,3}t}\, dt + N_2(0)$

where again the integral is from 0 to t. Separating the terms in the integral:

(3.55) $N_2(t)e^{\lambda_{2,3}t} = \int(\lambda_{1,2}R_{in}\, e^{\lambda_{2,3}t} / \lambda_{1,2})dt - \int(\lambda_{1,2}R_{in}\, e^{-\lambda_{1,2}t}\, e^{\lambda_{2,3}t} / \lambda_{1,2})dt$

$+ \int(\lambda_{1,2}N_1(0)\, e^{-\lambda_{1,2}t})\, e^{\lambda_{2,3}t}\, dt + N_2(0)$

$= (\lambda_{1,2}R_{in}/ \lambda_{1,2}) \int e^{\lambda_{2,3}t}dt - (\lambda_{1,2}R_{in} / \lambda_{1,2}) \int e^{-\lambda_{1,2}t}\, e^{\lambda_{2,3}t}\, dt$

$+ \lambda_{1,2}N_1(0) \int e^{-\lambda_{1,2}t}\, e^{\lambda_{2,3}t}\, dt + N_2(0)$

Noting that $e^a e^b = e^{(a+b)}$, Equation3.55 may be written further as:

(3.56) $N_2(t)e^{\lambda_{2,3}t} = (\lambda_{1,2}R_{in}/ \lambda_{1,2}) \int e^{\lambda_{2,3}t}dt - (\lambda_{1,2}R_{in} / \lambda_{1,2}) \int e^{\lambda_{2,3}t-\lambda_{1,2}t}\, dt$

$+ \lambda_{1,2}N_1(0) \int e^{\lambda_{2,3}t-\lambda_{1,2}t}\, dt + N_2(0)$

or

(3.57) $N_2(t)e^{\lambda_{2,3}t} = [(\lambda_{1,2}R_{in}/ \lambda_{1,2})\, (e^{\lambda_{2,3}t} - e^0) / \lambda_{2,3}]$

$- [(\lambda_{1,2}R_{in} / \lambda_{1,2})\, (e^{\lambda_{2,3}t-\lambda_{1,2}t} - e^0) / (\lambda_{2,3}-\lambda_{1,2})]$

$+ [\lambda_{1,2}N_1(0)\, (e^{\lambda_{2,3}t-\lambda_{1,2}t} - e^0) / (\lambda_{2,3}-\lambda_{1,2})] + N_2(0)$

Finally, dividing both sides of Equation 3.57 by $e^{\lambda_{2,3}t}$ and simplifying produces:

(3.58) $N_2(t) = [R_{in}(1 - e^{-\lambda_{2,3}t}) / \lambda_{2,3}]$

$- [R_{in}\, (e^{-\lambda_{1,2}t} - e^{-\lambda_{2,3}t}) / (\lambda_{2,3}-\lambda_{1,2})]$

$+ \lambda_{1,2}N_1(0)\, (e^{-\lambda_{1,2}t} - e^{-\lambda_{2,3}t}) / (\lambda_{2,3}-\lambda_{1,2})] + N_2(0)\, e^{-\lambda_{2,3}t}$

The solution for $N_3(t)$ then follows in the same manner as the solution for $N_2(t)$ with Equation 3.58 inserted into Equation 3.50 and solved using Bernoulli's method. Figure 3.11 shows the results for the first two compartments of a chain using Equations 3.51 and 3.58 for the case where all initial values are 0. Note the rise of the first compartment to equilibrium, followed later by the equilibrium of the second compartment. Note also the different curvatures of the solutions to the two compartments.

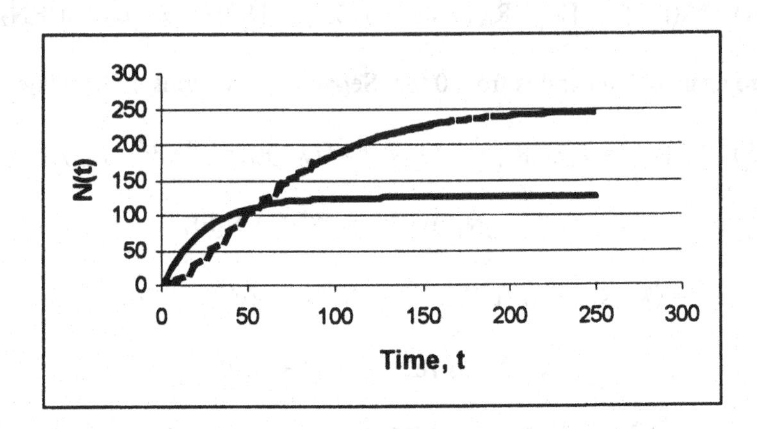

Figure 3.11. Example solutions to the first two compartments of a catenary chain. The solid line is for the first compartment and the dashed line is for the second. The rate into the first compartment is 5 g/d; the rate constant from the first to the second compartments is 0.04 d^{-1}; and the rate constant out of the second compartment is 0.02 d^{-1}. Time on the x-axis is in days and N(t) is in grams.

Bernoulli's method can be used for all compartments further down a catenary chain. It should be noted, however, that the equation for $N_2(t)$ is significantly larger than the equation for $N_1(t)$, and the equation for $N_3(t)$ will be larger than the equation for $N_2(t)$. This pattern holds in general: as compartments further down the chain are considered, the integral in Equation 3.15 becomes larger. Still, the same kinds of integrals encountered in solving for $N_1(t)$ and $N_2(t)$ appear, and so new mathematical tools are needed. The task of solution simply becomes more laborious (and tedious!).

3.5.7. Non-constant parameters

All of the models in Sections 3.5.1 through 3.5.6 were based on the assumption that transfer or transformation was either zeroth-order or first-order. The transfer rate constants literally were constant in time (i.e. λ and k did not change in time). It is possible, however, for these values to change with time in some systems. Consider, for example, the transfer of particles from the atmosphere to soil. The rate of transfer depends on the size of the particles, with heavier particles sedimenting more rapidly and very small particles kept aloft by wind. As time passes, the size distribution of the particles will change since only the particles with low settling characteristics will survive in the air. The result will be an apparent change in any rate constants for sedimentation and transfer.

Such situations may still be solved with Bernoulli's method, although only if the dependence of λ and k on time is not too complicated. Returning to the simple case of exponential washout in Section 2.5.2, we will consider material or energy placed into a compartment at t equal 0. This initial amount is N(0). There is no further inflow into the compartment over time. The material or energy is removed from the compartment by a first order process, but now the transfer rate constant is a function of time, $\lambda(t)$. The differential equation for this compartment then is:

(3.59) $$dN(t)/dt = 0 - \lambda(t)\, N(t)$$

the solution to which, following the derivation in Section 3.5.2, is:

(3.60) $$N(t) = N(0)e^{-a}$$

where

(3.61) $$a = \int \lambda(t)dt$$

and the integration is from 0 to t. An analytic solution to N(t) for the case of simple washout may then be found for any functions $\lambda(t)$ possessing a solution to the integral in Equation 3.61.

For example, let $\lambda(t)$ equal ct (λ increasing linearly in time). Placing this function for $\lambda(t)$ into Equation 3.61 yields:

(3.62) $$a = \int \lambda(t)dt = \int ct\, dt = c\int t\, dt = ct^2/2$$

which may be inserted into Equation 3.61 to yield:

(3.63) $$N(t) = N(0)\exp(-ct^2/2)$$

where exp is the exponential function (this symbol for the exponential function was chosen here to make it easier to see the exponent, which involves t to the second power).

As another example, let $\lambda(t)$ equal ce^{-kt} (λ decreasing in time). Placing this function for $\lambda(t)$ into Equation 3.61 yields:

$$(3.64) \qquad a = \int \lambda(t)dt = \int ce^{-kt}\,dt = c\int e^{-kt}\,dt = c(1-e^{-kt})/k$$

which may be inserted into Equation 3.60 to yield:

$$(3.65) \qquad N(t) = N(0)\exp(-c(1-e^{-kt})/k)$$

where exp is the exponential function.

Similar analytic solutions to N(t) may be found whenever a closed form solution exists to the integral in Equation 3.62. Significantly greater problems arise when cases more complex than simple washout are modeled. Returning to Equation 3.15, note that in the two examples above Q(x) was zero because there was no inflow into the compartment. In general, this may not be a valid assumption and Q(x) may have a non-zero value or even be a function of time. In such cases, it will be necessary to integrate the product $Q(x) \bullet e^{a}$. As $\lambda(t)$ or k(t) become more complicated functions, one soon runs into the problem of being unable to find an analytic solution to the integral of $Q(x) \bullet e^{a}$. Bernoulli's method falls apart in these cases when analytic solutions are sought, although numerical solutions to the integrals may still be obtained.

3.5.8. Systems with transfer and transformation

Finally, consider problems in which a pollutant transfers between compartments of a system, and also transforms within the compartments. Figure 3.12 shows such a problem, with three compartments (air, soil and water) linked into a chain. In addition, the pollutant is transformed chemically in each of the compartments, with the transformation resulting in three chemical forms shown as A, B and C.

The process for solution in this case may proceed in several ways, but clearly there are limits on the order in which the differential equations must be solved. It may be noted that the solution to the compartment containing chemical A in air does not depend on the solutions to any of the other compartments. One must, therefore, begin with the solution for chemical A in air. The second step may involve either the solution to chemical A in soil (which involves only the solution to chemical A in air) or chemical B in air (which also involves only the solution to chemical A in air. The next set of steps involves either the solution to chemical A in water (which involves only the solution to chemical A in soil), or chemical B in soil (which involves the solution to chemical A in soil and chemical B in air), or chemical C in air (which involves only the solution to chemical B in air). In general, the sequence of solutions proceeds along the diagonals beginning with the diagonal in the upper left and proceeding to the diagonal in the lower right of the figure. This ensures one does not encounter a differential equation that cannot be solved with the methods discussed in this chapter.

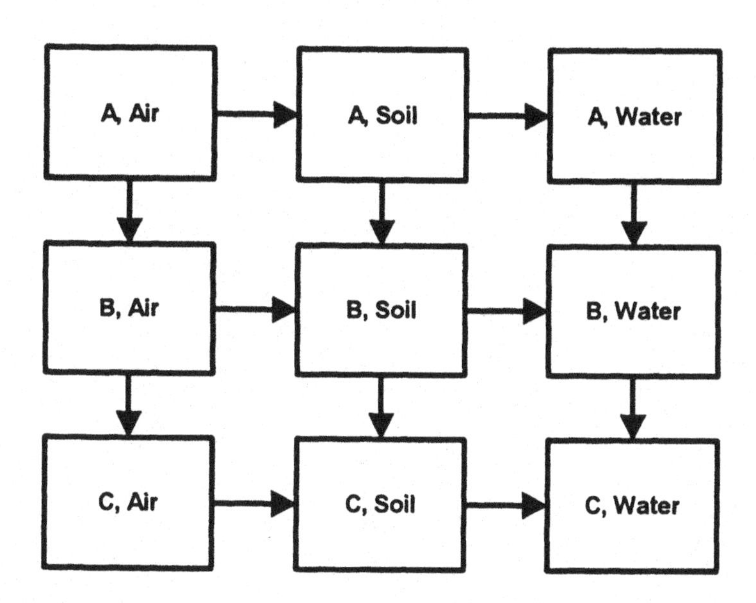

Figure 3.12. A series of compartments involving transfer (from air to soil to water) and transformation (from form A to B to C).

References

1. D. Crawford-Brown, *Theoretical and Mathematical Foundations of Human Health Risk Analysis*, Kluwer Academic Publishers, Boston, 1997.
2. R. Smith and T. Smith, *Elements of Ecology*, Benjamin Cummings, Menlo Park, CA, 1998.
3. D. Crawford-Brown, *Risk-Based Environmental Decisions: Methods and Culture*, Kluwer Academic Publishers, Boston, 2000.
4. L. Kells, *Elementary Differential Equations*, McGraw-Hill Book Company, Inc., New York, 1960.
5. M. Allaby, *Basics of Environmental Science*, Routledge, London, 1996.
6. I. Gradshteyn and I. Ryzhik, *Table of Integrals, Series and Products*, Academic Press, Orlando, FL, 1970.

CHAPTER 4
Laplace Transforms and Coupled Differential Equations

4.1. Coupled Systems and Feedback

Chapter 3 considered systems in which the rate of flow between two compartments was described by zeroth or first order kinetics, and where flow was in one direction only. Solutions were obtained by solving for the amount in the first compartment of the chain, and then proceeding through the chain to the final compartment in the order in which compartments are encountered. This process of solution was effective because Bernoulli's solution to any one compartment involved knowledge only of the functions describing compartments "higher" in the chain. At no time did the differential equation for a given compartment involve information on compartments "further down" the chain.

While the evolution of the state vector for some environmental systems can be approximated by such uncoupled systems of differential equations, most environmental systems have flow in both directions between compartments. At the least, flow is in a *cycle* [1], perhaps moving through a chain of compartments but eventually arriving back at the original compartment. Pesticide spilled into the water of a lake settles to the sediment, is taken up by reeds and, when the reeds die and decay, passes back into the water. Carbon entering the air may be taken up by plants, move into the soil when the plants die and rot, be eroded into the ocean, and travel into the interior of the earth during subduction. Eventually, it enters back into the air as tectonic activity brings carbon dioxide to the surface. The major nutrients (nitrogen, phosphorus, etc) move through cycles. Such cycles of material and energy make up the bulk of environmental processes. For such processes, methods used to solve uncoupled systems of differential equations will not suffice, since the solution to each compartment depends on the solutions to all of the others. The compartments of such systems, and the differential equations describing them, are *coupled*.

To complicate matters, parameter values in these coupled equations may also be functions of the state vector for the system. As many systems evolve, signals are transferred back to the compartments that affect the rates in, the rates out, or both. These signals constitute *feedback* [2] and have the effect of either bringing the state vector back to some initial value or causing the state vector to spiral out of control. If the effect is to bring the system back to an initial state vector, the feedback is said to be *negative*. If the effect is to cause the state vector to evolve more rapidly, the feedback is said to be *positive*. Regardless of whether feedback is positive or negative, the result usually is a differential equation that cannot be solved by methods applicable only to uncoupled systems.

The carbon cycle between the atmosphere and flora (plants) provides a simple example of feedback. Consider the issue of global warming, in which temperature may depend on the amount of carbon (in the form of carbon dioxide molecules) in the atmosphere. As the amount of carbon in the atmosphere increases, plants may be stimulated to grow. As the plants grow, they increase the rate at which carbon transfers out of the atmosphere. This, in turn, decreases the rate of increase of carbon in the atmosphere, pulling the atmospheric carbon back to an initial value. The increase in atmospheric carbon has produced a signal from the plants (increased absorption of carbon dioxide) that has a negative effect on the rate of growth of atmospheric carbon. The feedback in this case is negative.

If, however, the increase in atmospheric carbon were to produce an increase in temperature leading to drought, plants might die. In that case, the rate of flow of carbon from the atmosphere to the flora would decrease, allowing the atmospheric carbon to build up more rapidly. As the atmospheric carbon built up to higher levels, more plants would die, leading to an even lower rate of transfer from atmosphere to flora, and so on. The feedback in this case is positive.

Negative feedback in this example represents *homeostasis* [3], or the tendency of a controlled system to move back to its original condition. Homeostatic mechanisms prevent the state vector for a system from spiraling out of control; these mechanisms adjust the system parameters so the original state is restored. Small initial changes in a state vector (such as the introduction of an amount of carbon into the atmosphere, perhaps by burning) are referred to as *perturbations*. If a homeostatic mechanism is present, this perturbation causes a change in the system that in turn decreases the effect of the perturbation. In the example from the carbon cycle, the plants provided a homeostatic mechanism that brought the atmospheric carbon back to its initial value by increasing the rate of transfer from the atmosphere to the flora whenever the amount of atmospheric carbon increased.

Many environmental systems contain homeostatic mechanisms that operate within some limits of perturbations. The plants might, for example, respond to an increase in atmospheric carbon by growing at a faster rate, increasing the transfer of carbon from atmosphere to flora, and decreasing the amount of atmospheric carbon. Similarly, if atmospheric carbon decreased below an initial value, there would be less plant growth, the rate of transfer from atmosphere to flora would decrease, and atmospheric carbon would increase back to its initial value. This homeostatic mechanism could maintain the atmospheric carbon at initial values so long as the perturbations were not too large. If, however, the amount of atmospheric carbon were suddenly to decrease by a large amount, the temperature might plunge and the plants would be killed. This would essentially break the homeostatic mechanism. Where homeostatic mechanisms exist, environmental management focuses on the goal of preventing perturbations to the system that are so large as to break the mechanisms. So long as the perturbations are small, the homeostatic mechanisms might be counted on to restore the state of the environment.

This chapter considers models of coupled systems of environmental compartments, with feedback between those compartments. Bernoulli's method does not work in such cases, since the solution to the differential equation for each compartment depends on the solutions to the other compartments. It always is possible to solve such coupled systems of equations numerically, as will be done in Chapter 7, but this chapter considers only analytic solutions. The particular technique used here is Laplace transforms.

4.2. Transforms

Transforms represent a broad class of mathematical techniques that do exactly what the name implies: they transform a mathematical problem into a different set of conditions. They are similar to translation in language. They change (transform) the structure of a problem into a new structure (or language) through a process of transformation (not to be confused with the processes of chemical, physical and biological transformation). The problem is solved in this new structure (this new language) and the solution obtained in the new structure is translated back into the original structure through a process of *inverse transformation*.

Why bother to transform a problem? Generally, transforms are applied when it is difficult to find a solution to a mathematical problem under the original structure. Finding the solution to coupled systems of differential equations in the language of differential equations is difficult at best and often impossible due to the coupling. Laplace transforms [4], however, transform the coupled system of differential equations into a system of algebraic equations that may then be solved using standard techniques of algebra. Having found the solutions in the transformed system, it might then be possible to translate the solutions back into the original language of the problem, yielding the solutions to the original coupled system of differential equations. The tasks involved in the use of Laplace transforms for solving coupled systems of differential equations then are:

- Transform the original system of differential equations into a system of algebraic expressions in the language of Laplace transforms.

- Solve the system of algebraic equations to obtain the state vector in the language of Laplace transforms.

- Translate the state vector (now in the language of Laplace transforms) back into the original language to obtain the desired state vector for the original system.

If any of these three tasks fail, the process will not suffice. Some transforms work better than others in specific cases because the three tasks above are simpler. It is not unusual to try a transform technique only to find it floundering at one or more of the tasks, requiring that a different technique be tried. Since a large class of simple coupled environmental systems can be solved using Laplace transforms, that method is employed here. More complex systems are treated in Chapter 7 through numerical solution of differential equations.

Note: Transform techniques, including Laplace transforms, work well for relatively simple systems of a few compartments. While there are important problems in environmental risk assessment that satisfy this condition, it also is the case that many problems involve significantly more compartments, with complex patterns of feedback. In fact, it is surprising how rapidly the complexity of an environmental model can grown beyond the capabilities of any method for obtaining analytic solutions. Be prepared to give in at times and move to numerical methods!

Since this is not a text in pure mathematics, but rather a book of mathematics applied to modeling, this chapter does not delve into the reasons Laplace transforms work as they do. The focus instead is on how Laplace transforms are applied to coupled differential equations in generating solutions. We will consider the example shown in Figure 4.1 throughout the chapter, and various simplifications of this system. To vary the examples from those used in earlier chapters, we will consider here a problem of *pharmacokinetics*, or transfer of a pollutant between compartments representing organs or tissues in the human body [5]. Such models are defined by the same conservation of mass principles used in developing the differential equations for environmental systems. The state vector for such a system then describes the amount (or concentration) of a pollutant in each of the tissues or organs of the body.

In this example, a pollutant (such as the pesticide atrazine) moves into the body through a route of intake. This might be through ingestion, so the pesticide enters through the gastrointestinal (GI) tract. It is assumed to enter immediately into the bloodstream, from which it is transferred by a series of first-order processes into various organs and tissues. From these organs and tissues, there is transfer either back to the bloodstream or out of the body through excretion in the urine or feces. The task is to find the amount of the pesticide in each of the compartments of the body as a function of time.

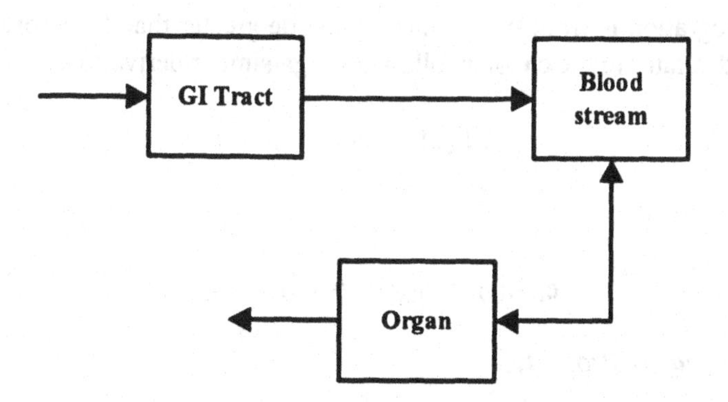

Figure 4.1. An example of a coupled system of differential equations. This example displays the transfer of a pollutant between some of the compartments of the human body, and is used throughout this chapter. Note that the flow from the bloodstream to the organ is bi-directional, precluding the use of Bernoulli's method.

4.3. The Laplace Transform

Consider a function, N(t) describing the amount of pesticide in one of the compartments of Figure 4.1 as a function of time, t. The Laplace transform of N(t) will be shown as £{N(t)} and is defined by [4]:

(4.1) $$£\{N(t)\} = n(s) = \int e^{-st} N(t)\, dt$$

where the integral is from 0 to ∞. Note that N(t) is replaced by n(s) when the Laplace transform is applied to N(t). This convention of replacing the function name (here, N) by the lower case version of the same letter, and replacing the argument (here, t) by the symbol s will be employed for all functions. For example, $£\{G(t)\} = g(s)$; $£\{N_i(t)\} = n_i(s)$; $£\{N(x)\} = n(s)$; etc.

The solutions to differential equations considered in this chapter will require that the Laplace transform be taken of various combinations of parameters and functions. These are reviewed in the theorems below. As an example of the development of the Laplace transform, consider the case in which N(t) equals 1. Using Equation 4.1:

Theorem 1.1 $£\{N(t)\} = £\{1\} = \int e^{-st} \bullet 1\, dt = \int e^{-st}\, dt = (e^{-0} - e^{-\infty}) / s = 1/s$

Where the integration is from 0 to ∞ and s must be greater than 0. A series of theorems needed later in the chapter follow through similar derivations.

Theorem 1.2 $$£\{c\} = c/s$$

where c is a constant.

Theorem 1.3 $$£\{c_1N_1(t)+c_2N_2(t)\} = c_1n_1(s)+c_2n_2(s)$$

which is the *Linearity Property*.

Theorem 1.4 $$£\{e^{at}N(t)\} = n(s-a)$$

which is the *Shifting Property*.

Theorem 1.5 $$£\{N(ct)\} = n(s/c)/c$$

which is the *Change of Scale Property*.

There then is a series of theorems related to the Laplace transform of derivatives, which will be needed in solving differential equations.

Theorem 1.6 $$£\{dN(t)/dt\} = sn(s) - N(0)$$

where N(0) is the initial value of the function N(t). Higher order derivatives may be found from the general relation:

Theorem 1.7 $£\{N^i(t)\} = s^i n(s) - s^{i-1}N(0) - s^{i-2}N^1(0) \ldots s^1N^{i-2}(0) - N^{i-1}(0)$

where $N^i(t)$ indicates in the ith derivative of N(t), or the order of the term. Note that there are i+1 terms on the right hand side of this relationship, and that it is necessary to find a series of derivatives of N(t) and to evaluate these at t=0. In other words, the solution to the differential equation may involve knowledge not only of the initial value of the function, but also the initial value of the derivatives of the function. Theorem 1.6 can be seen to follow directly from Theorem 1.7 by noting that i equals 1 (so only the first and last terms in Theorem 1.7 exist).

Very large compilations of other theorems exist [6] and should be consulted for more complex functions of N(t). The theorems above will suffice for the systems of first-order differential equations considered in this chapter.

Example 4.1. Consider the differential equation $dN(t)/dt = 3 - 0.1N(t)$, where N(0) is 4. Using Theorem 1.3:

$$£\{dN(t)/dt\} = £\{3\} - £\{0.1N(t)\}$$

Using Theorem 1.7 for the left hand side:

$$£\{dN(t)/dt\} = sn(s) - N(0)$$

Using Theorems 1.1 and 1.2 for the first term on the right hand side:

$$£\{3\} = 3/s$$

Using Theorem 1.3 for the second term on the right hand side:

$$£\{0.1N(t)\} = 0.1n(s)$$

so the original differential equation has been transformed to:

$$sn(s) - N(0) = 3/s - 0.1n(s)$$

4.4. The Inverse Laplace Transform

If n(s) is the Laplace transform of the function N(t), then N(t) is the *inverse Laplace transform* of n(s). We will show the inverse Laplace transform as $£^{-1}\{n(s)\} = N(t)$. Again, there are compilations of a large number of inverse

Laplace transforms [4], and readers should consult these for the many special cases likely to arise in the use of Laplace transforms. For the problems encountered in this chapter, the relationship of most use is the *Heaviside Expansion Theorem* [4]. Let P(s) and Q(s) be two polynomials. The degree of P(s), or power of the highest order term, must be less than the degree of Q(s). Q(s) has m distinct zeroes, or values of s for which Q(s) is zero. We will call these α_k, where k is from 1 to m. If these conditions apply (and they will for the models solved in this chapter), then:

Theorem 1.8 $$£^{-1}\{P(s)/Q(s)\} = \Sigma\, [\, P(\alpha_k) / Q^1(\alpha_k)]\, e^{\alpha_k t}$$

where the summation is over all m values of k (i.e. from k=1 to k=m) and where $Q^1(\alpha_k)$ indicates the first derivative of Q(s) evaluated at the value α_k. In addition, the following Linearity Property of inverse Laplace transforms will be needed:

Theorem 1.9 $$£^{-1}\{c_1 n_1(s)+c_2 n_2(s)\} = c_1 £^{-1}\{n_1(s)\} + c_2 £^{-1}\{n_2(s)\} = c_1 N_1(t) + c_2 N_2(t)$$

Note that the solution to Example 4.2 using Laplace transforms is the same as that obtained using Bernoulli's method. This may be seen by comparing the solution above to Equation 3.34. Either Laplace transforms or Bernoulli's method may be used for this example since there is a single differential equation and no coupling. Whenever two methods may be applied in obtaining a solution, the answer should be the same from those two methods.

4.5. Applications of Laplace Transforms

This section presents a series of applications of Laplace transforms, using problems of increasing complexity. In all cases, the examples are special cases of the problem shown in Figure 4.1.

4.5.1. Partitioning between compartments in closed systems

Figure 4.2 shows a simplified version of Figure 4.1. The system has been reduced to the bloodstream and an organ with which a pesticide is exchanged. It is a closed system because the pesticide transfers between the two organs but never leaves the system of two organs. At t equal 0, an initial quantity, $N_b(0)$ of the pesticide is placed into the bloodstream, and an initial quantity, $N_o(0)$ of the pesticide is placed into the organ. The subscript b refers to the blood; the subscript o refers to the organ. The pesticide then transfers from the blood to the organ with a first-order transfer rate constant λ_{bo}, and from the organ back to the blood with a first-order transfer rate constant λ_{ob}.

Example 4.2. Consider Example 4.1 which yielded: $sn(s) - N(0) = 3/s - 0.1n(s)$

Rearranging terms yields:

$(s+0.1)n(s) = N(0) + 3/s$ or

$n(s) = [N(0) + 3/s]/(s+0.1) = N(0)/(s+0.1) + 3/[s(s+0.1)]$

Applying Theorem 1.9:

$£^{-1}\{N(0)/(s+0.1) + 3/[s(s+0.1)]\} = N(0)£^{-1}\{1/(s+0.1)\} + 3£^{-1}\{1/(s(s+0.1))\}$

Using the Heaviside expansion theorem, and considering the first term on the right hand side, note that P(s) is 1; Q(s) is (s+0.1); $Q^1(s)$ is 1; and the root of Q(s) is –0.1. Therefore:

$N(0)£^{-1}\{1/(s+0.1)\} = N(0)(1/1)e^{-0.1t} = N(0)e^{-0.1t}$

For the second term on the right hand side, P(s) is 1; Q(s) is s(s+0.1); $Q^1(s)$ is (2s+0.1); and there are two roots of Q(s): 0 and –0.1. Therefore:

$3£^{-1}\{1/(s(s+0.1))\} = 3(1/(2\cdot 0+0.1))e^{0\cdot t} + 3(1/(2\cdot(-0.1)+0.1))e^{-0.1\cdot t}$

Finally:

$£^{-1}\{n(s)\} = N(t) = N(0)e^{-0.1t} + 30e^{0\cdot t} - 30e^{-0.1\cdot t} = N(0)e^{-0.1t} + 30(1 - e^{-0.1\cdot t})$

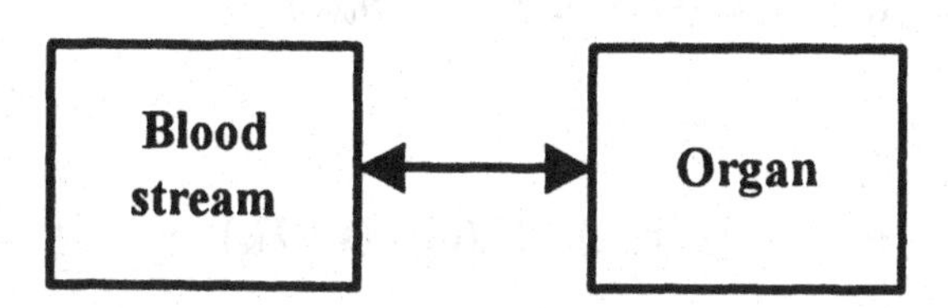

Figure 4.2. A simplified version of Figure 4.1, with bi-directional flow between two compartments. This example is used to demonstrate the principle of partitioning between compartments in a closed system.

The two differential equations for this system (one for each compartment) are given by:

(4.2) $$dN_b(t)/dt = \lambda_{ob}N_o(t) - \lambda_{bo}N_b(t)$$

(4.3) $$dN_o(t)/dt = \lambda_{bo}N_b(t) - \lambda_{ob}N_o(t)$$

Note that this system of equations cannot be solved using Bernoulli's method since the solution to the first equation would require the solution to the second, and the solution to the second would require the solution to the first. Laplace transforms, however, will allow us to (i) transform this system of differential equations into a system of algebraic expressions involving $n_b(s)$ and $n_o(s)$, (ii) solve this algebraic system, and then (iii) apply the inverse Laplace transform to convert the algebraic solution back to the solutions for $N_b(t)$ and $N_o(t)$.

Applying Theorem 1.3 to both equations, the following relations are obtained:

(4.4) $$\pounds\{dN_b(t)/dt\} = \pounds\{\lambda_{ob}N_o(t)\} - \pounds\{\lambda_{bo}N_b(t)\}$$

(4.5) $$\pounds\{dN_o(t)/dt\} = \pounds\{\lambda_{bo}N_b(t)\} - \pounds\{\lambda_{ob}N_o(t)\}$$

Using Theorem 1.7 for the left hand sides and Theorem 1.3 again for the right hand sides yields:

(4.6) $$sn_b(s) - N_b(0) = \lambda_{ob}n_o(s) - \lambda_{bo}n_b(s)$$

(4.7) $$sn_o(s) - N_o(0) = \lambda_{bo}n_b(s) - \lambda_{ob}n_o(s)$$

Using Equation 4.6 to solve for $n_b(s)$:

(4.8) $$(s + \lambda_{bo})n_b(s) = \lambda_{ob}n_o(s) + N_b(0)$$

or

(4.9) $$n_b(s) = [\lambda_{ob}n_o(s) + N_b(0)] / (s + \lambda_{bo})$$

Substituting Equation 4.9 into Equation 4.7 yields:

(4.10) $$sn_o(s) - N_o(0) = \lambda_{bo}[\lambda_{ob}n_o(s) + N_b(0)] / (s + \lambda_{bo}) - \lambda_{ob}n_o(s)$$

Collecting terms involving $n_o(s)$:

(4.11) $[s + \lambda_{ob} - \lambda_{bo}(\lambda_{ob}/ (s + \lambda_{bo}))]n_o(s) = \lambda_{bo}N_b(0)/ (s + \lambda_{bo}) + N_o(0)$

Multiplying both sides by $(s+\lambda_{bo})$ yields:

(4.12) $$[(s + \lambda_{ob})(s + \lambda_{bo}) - \lambda_{bo}(\lambda_{ob})]n_o(s) = \lambda_{bo}N_b(0) + N_o(0)(s + \lambda_{bo})$$

or

(4.13) $$n_o(s) = [\lambda_{bo}N_b(0) + N_o(0)(s + \lambda_{bo})]/[(s + \lambda_{ob})(s + \lambda_{bo}) - \lambda_{bo}\lambda_{ob}]$$

$N_o(t)$ may then be obtained by first applying the inverse Laplace transform to both sides of the equation:

(4.14) $$N_o(t) = \pounds^{-1}\{n_o(s)\} =$$

$$\pounds^{-1}\{[\lambda_{bo}N_b(0) + N_o(0)(s + \lambda_{bo})]/[(s + \lambda_{ob})(s + \lambda_{bo}) - \lambda_{bo}\lambda_{ob}]\}$$

Using Theorem 1.9:

(4.15) $$N_o(t) = \pounds^{-1}\{\lambda_{bo}N_b(0)/[(s + \lambda_{ob})(s + \lambda_{bo}) - \lambda_{bo}\lambda_{ob}]\}$$

$$+ \pounds^{-1}\{N_o(0)(s + \lambda_{bo})/[(s + \lambda_{ob})(s + \lambda_{bo}) - \lambda_{bo}\lambda_{ob}]\}$$

The right hand side may be solved using the Heaviside Expansion Theorem applied twice (once to each of the two terms). For the first term:

(4.16) $$P(s) = \lambda_{bo}N_b(0)$$

and

(4.17) $$Q(s) = [(s + \lambda_{ob})(s + \lambda_{bo}) - \lambda_{bo}\lambda_{ob}] = s^2 + s\lambda_{bo} + s\lambda_{ob} + \lambda_{ob}\lambda_{bo} - \lambda_{ob}\lambda_{bo}$$

$$= s^2 + s\lambda_{bo} + s\lambda_{ob} = s(s + (\lambda_{bo} + \lambda_{ob}))$$

and

(4.18) $$Q^1(s) = 2s + \lambda_{bo} + \lambda_{ob}$$

The roots of Equation 4.17 are $\alpha_1 = 0$ and $\alpha_2 = -(\lambda_{bo} + \lambda_{ob})$. Using Theorem 1.8:

(4.19) $\pounds^{-1}\{\lambda_{bo}N_b(0)/[(s + \lambda_{ob})(s + \lambda_{bo}) - \lambda_{bo}\lambda_{ob}]\}$

$$= [\lambda_{bo}N_b(0)/(2\bullet 0 + \lambda_{bo} + \lambda_{ob})]\exp[0t]$$

$$+[\lambda_{bo}N_b(0)/(-2(\lambda_{bo} + \lambda_{ob}) + \lambda_{bo} + \lambda_{ob}]\exp[-(\lambda_{bo} + \lambda_{ob})t]$$

$$= [\lambda_{bo}N_b(0)/(\lambda_{bo} + \lambda_{ob})] - [\lambda_{bo}N_b(0)/(\lambda_{bo} + \lambda_{ob})]\exp[-(\lambda_{bo} + \lambda_{ob})t]$$

$$= [\lambda_{bo}N_b(0)/(\lambda_{bo} + \lambda_{ob})](1 - \exp[-(\lambda_{bo} + \lambda_{ob})t])$$

Note: the exponential function is shown in this relation as exp[x] rather than e^x to make the subscripts on λ clear.

For the second term on the right hand side of Equation 4.15:

(4.20) $$P(s) = N_o(0)(s + \lambda_{bo})$$

and

(4.21) $$Q(s) = [(s + \lambda_{ob})(s + \lambda_{bo}) - \lambda_{bo}\lambda_{ob}] = s^2 + s\lambda_{bo} + s\lambda_{ob} + \lambda_{ob}\lambda_{bo} - \lambda_{ob}\lambda_{bo}$$

$$= s^2 + s\lambda_{bo} + s\lambda_{ob} = s(s + (\lambda_{bo} + \lambda_{ob}))$$

and

(4.22) $$Q^1(s) = 2s + \lambda_{bo} + \lambda_{ob}$$

The roots of Equation 4.21 are $\alpha_1 = 0$ and $\alpha_2 = -(\lambda_{bo} + \lambda_{ob})$. Using Theorem 1.8:

(4.23) $$£^{-1}\{N_o(0)(s+\lambda_{bo})/[(s + \lambda_{ob})(s + \lambda_{bo}) - \lambda_{bo}\lambda_{ob}]\}$$

$$= [N_o(0)(0+\lambda_{bo})/(2\bullet 0 + \lambda_{bo} + \lambda_{ob})]\exp[0t]$$

$$+[N_o(0)(-(\lambda_{bo}+\lambda_{ob})+\lambda_{bo}))/(-2(\lambda_{bo}+\lambda_{ob}) + \lambda_{bo} + \lambda_{ob})]\exp[-(\lambda_{bo} + \lambda_{ob})t]$$

$$= [N_o(0)\lambda_{bo}/(\lambda_{bo} + \lambda_{ob})]$$

$$+ [N_o(0)\lambda_{ob}/(\lambda_{bo} + \lambda_{ob})]\exp[-(\lambda_{bo} + \lambda_{ob})t]$$

Combining Equations 4.19 and 4.23:

(4.24) $$N_o(t) = [\lambda_{bo}N_b(0)/(\lambda_{bo} + \lambda_{ob})](1 - \exp[-(\lambda_{bo} + \lambda_{ob})t])$$

$+ [N_0(0)\lambda_{bo}/(\lambda_{bo} + \lambda_{ob})] + [N_0(0)\lambda_{ob}/(\lambda_{bo} + \lambda_{ob})]\exp[-(\lambda_{bo} + \lambda_{ob})t]$

Note: again, the exponential function is shown in this relation as exp[x] rather than e^x to make the subscripts on λ clear.

There are now 2 ways to solve for $N_b(t)$. The solution for $N_o(t)$ might be substituted back into Equation 4.2. This would yield a differential equation for $N_b(t)$ involving only $N_b(t)$. Bernoulli's method might then be applied. An alternative is to substitute Equation 4.13 into Equation 4.6 and continue using Laplace transform methods. The latter approach provides additional practice at the use of Laplace transforms, and so is adopted here. Making the indicated substitution:

(4.25) $sn_b(s) - N_b(0) =$

$$\lambda_{ob}[\lambda_{bo}N_b(0) + N_o(0)(s + \lambda_{bo})]/[(s + \lambda_{ob})(s + \lambda_{bo}) - \lambda_{bo}\lambda_{ob}] - \lambda_{bo}n_b(s)$$

Collecting terms involving $n_b(s)$:

(4.26) $(s + \lambda_{bo})n_b(s) =$

$$N_b(0) + \lambda_{ob}[\lambda_{bo}N_b(0) + N_o(0)(s + \lambda_{bo})]/[(s + \lambda_{ob})(s + \lambda_{bo}) - \lambda_{bo}\lambda_{ob}]$$

Dividing both sides by $(s + \lambda_{bo})$ yields:

(4.27) $n_b(s) =$

$$[N_b(0)/(s + \lambda_{bo})] + \lambda_{ob}\lambda_{bo}N_b(0)/[(s + \lambda_{bo})[(s + \lambda_{ob})(s + \lambda_{bo}) - \lambda_{bo}\lambda_{ob}]]$$

$$+ \lambda_{ob}N_o(0)/[(s + \lambda_{ob})(s + \lambda_{bo}) - \lambda_{bo}\lambda_{ob}]$$

This solution involves three terms. Applying the inverse Laplace transform to all terms using Theorem 1.9:

(4.28) $N_b(t) = \pounds^{-1}\{N_b(0)/(s + \lambda_{bo})\}$

$$+ \pounds^{-1}\{\lambda_{ob}\lambda_{bo}N_b(0)/[(s + \lambda_{bo})[(s + \lambda_{ob})(s + \lambda_{bo}) - \lambda_{bo}\lambda_{ob}]]\}$$

$$+ \pounds^{-1}\{\lambda_{ob}N_o(0)/[(s + \lambda_{ob})(s + \lambda_{bo}) - \lambda_{bo}(\lambda_{ob})]\}$$

For the first term on the right hand side of Equation 4.28:

(4.29) $$P(s) = N_b(0)$$

and

(4.30) $$Q(s) = (s + \lambda_{bo})$$

and

(4.31) $$Q^1(s) = 1$$

The roots of Equation 4.30 are $\alpha_1 = -\lambda_{bo}$. Using Theorem 1.8:

(4.32) $$\pounds^{-1}\{N_b(0)/(s+\lambda_{bo})\} = [N_b(0)/1]\exp[-\lambda_{bo}t] = N_b(0)\exp[-\lambda_{bo}t]$$

For the second term on the right hand side of Equation 4.28:

(4.33) $$P(s) = \lambda_{ob}\lambda_{bo}N_b(0)$$

and

(4.34) $$Q(s) = (s + \lambda_{bo})[(s + \lambda_{ob})(s + \lambda_{bo}) - \lambda_{bo}\lambda_{ob}] =$$

$$= (s + \lambda_{bo})(s^2 + s\lambda_{bo} + s\lambda_{ob} + \lambda_{ob}\lambda_{bo} - \lambda_{ob}\lambda_{bo})$$

$$= (s + \lambda_{bo})(s^2 + s\lambda_{bo} + s\lambda_{ob}) = (s^3 + s^2(2\lambda_{bo} + \lambda_{ob}) + s(\lambda_{bo}\lambda_{bo} + \lambda_{bo}\lambda_{ob}))$$

$$= s(s + \lambda_{bo})(s + (\lambda_{bo} + \lambda_{ob}))$$

and

(4.35) $$Q^1(s) = [3s^2 + 2s(2\lambda_{bo} + \lambda_{ob}) + (\lambda_{bo}\lambda_{bo} + \lambda_{bo}\lambda_{ob})]$$

The roots of Equation 4.34 are $\alpha_1 = 0$; $\alpha_2 = -\lambda_{bo}$; and $\alpha_3 = -(\lambda_{bo} + \lambda_{ob})$. Using Theorem 1.8:

(4.36) $$\pounds^{-1}\{\lambda_{ob}\lambda_{bo}N_b(0)/[(s + \lambda_{bo})[(s + \lambda_{ob})(s + \lambda_{bo}) - \lambda_{bo}\lambda_{ob}]]\}$$

$$= [\lambda_{ob}\lambda_{bo}N_b(0)/[3\bullet 0^2 + 2\bullet 0\bullet(2\lambda_{bo} + \lambda_{ob}) + (\lambda_{bo}\lambda_{bo} + \lambda_{bo}\lambda_{ob})]]\exp[0t]$$

$$+ \lambda_{ob}\lambda_{bo}N_b(0)/[3(-\lambda_{bo})^2 + 2(-\lambda_{bo})(2\lambda_{bo} + \lambda_{ob}) + (\lambda_{bo}\lambda_{bo} + \lambda_{bo}\lambda_{ob})]\exp[-\lambda_{bo}t]$$

$+\lambda_{ob}\lambda_{bo}N_b(0)/[3(-(\lambda_{bo}+\lambda_{ob}))^2+2\bullet(-(\lambda_{bo}+\lambda_{ob}))(2\lambda_{bo}+\lambda_{ob})$

$+ (\lambda_{bo}\lambda_{bo}+\lambda_{bo}\lambda_{ob})]\exp[-(\lambda_{bo}+\lambda_{ob})t]$

$= \quad [\lambda_{ob}\lambda_{bo}N_b(0)/[(\lambda_{bo}\lambda_{bo} + \lambda_{bo}\lambda_{ob})]] - N_b(0)\exp[-\lambda_{bo}t]$

$+ \lambda_{ob}\lambda_{bo}N_b(0)/[\lambda_{ob}(\lambda_{bo}+\lambda_{ob})]\exp[-(\lambda_{bo}+\lambda_{ob})t]$

$= \quad [\lambda_{ob}N_b(0)/[(\lambda_{bo} + \lambda_{ob})]] - N_b(0)\exp[-\lambda_{bo}t]$

$+ [\lambda_{bo}N_b(0)/(\lambda_{bo}+\lambda_{ob})]\exp[-(\lambda_{bo}+\lambda_{ob})t]$

For the third term on the right hand side of Equation 4.28:

(4.37) $\quad P(s) = \lambda_{ob}N_o(0)$

and

(4.38) $\quad Q(s) = [(s + \lambda_{ob})(s + \lambda_{bo}) - \lambda_{bo}\lambda_{ob}] =$

$= (s^2 + s\lambda_{bo} + s\lambda_{ob} + \lambda_{ob}\lambda_{bo} - \lambda_{ob}\lambda_{bo})$

$= (s^2 + s\lambda_{bo} + s\lambda_{ob}) = s(s + (\lambda_{bo} + \lambda_{ob}))$

and

(4.39) $\quad Q^1(s) = [2s + (\lambda_{bo} + \lambda_{ob})]$

The roots of Equation 4.38 are $\alpha_1 = 0$ and $\alpha_2 = -(\lambda_{bo} + \lambda_{ob})$. Using Theorem 1.8:

(4.40) $\pounds^{-1}\{\lambda_{ob}N_o(0)/[(s + \lambda_{ob})(s + \lambda_{bo}) - \lambda_{bo}(\lambda_{ob})]\}$

$= [\lambda_{ob}N_o(0)/(2\bullet 0 + (\lambda_{bo} + \lambda_{ob}))]\exp[0t]$

$+ [\lambda_{ob}N_o(0)/(2\bullet(-(\lambda_{bo} + \lambda_{ob})) + (\lambda_{bo} + \lambda_{ob}))]\exp[-(\lambda_{bo} + \lambda_{ob})t]$

$= [\lambda_{ob}N_o(0)/(\lambda_{bo} + \lambda_{ob})]\exp[0t]$

$- [\lambda_{ob}N_o(0)/(\lambda_{bo} + \lambda_{ob})]\exp[-(\lambda_{bo} + \lambda_{ob})t]$

$$= [\lambda_{ob}N_o(0)/(\lambda_{bo} + \lambda_{ob})](1\text{-}\exp[-(\lambda_{bo} + \lambda_{ob})t])$$

Combining Equations 4.32, 4.36 and 4.40 yields:

$$(4.41)\quad N_b(t) = N_b(0)\exp[-\lambda_{bo}t] + [\lambda_{ob}N_b(0)/[(\lambda_{bo} + \lambda_{ob})]] - N_b(0)\exp[-\lambda_{bo}t]$$

$$+ [\lambda_{bo}N_b(0)/(\lambda_{bo}+\lambda_{ob})]\exp[-(\lambda_{bo}+\lambda_{ob})t] +$$

$$[\lambda_{ob}N_o(0)/(\lambda_{bo} + \lambda_{ob})](1\text{-}\exp[-(\lambda_{bo} + \lambda_{ob})t])$$

$$= \quad [\lambda_{ob}N_b(0)/[(\lambda_{bo} + \lambda_{ob})]] + [\lambda_{bo}N_b(0)/(\lambda_{bo}+\lambda_{ob})]\exp[-(\lambda_{bo}+\lambda_{ob})t] +$$

$$[\lambda_{ob}N_o(0)/(\lambda_{bo} + \lambda_{ob})](1\text{-}\exp[-(\lambda_{bo} + \lambda_{ob})t])$$

Note the similarity between Equations 4.24 and 4.41. Figure 4.3 shows a representative display of these two equations with all of the pesticide placed into the bloodstream at t equal 0, with λ_{bo} equal to 0.1 day^{-1} and λ_{ob} equal to 0.3 day^{-1}.

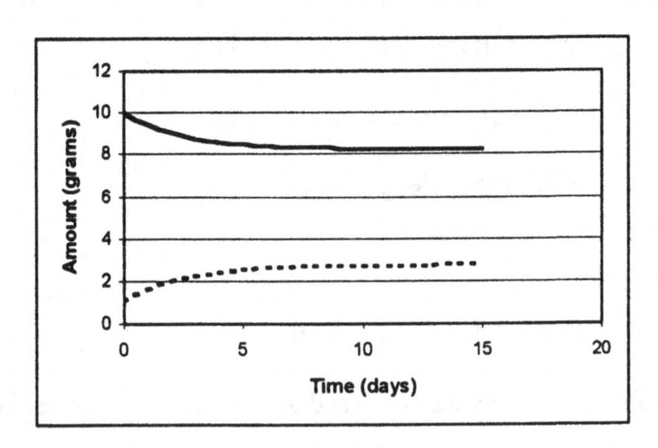

Figure 4.3. The amount of pesticide in the system shown in Figure 4.2 when the initial amount in the bloodstream is 10 grams, the initial amount in the organ is 1 gram, λ_{bo} is 0.1 day^{-1} and λ_{ob} is 0.3 day^{-1}. The solid line is the amount in the bloodstream and the dashed line is the amount in the organ. The equilibrium amounts in the bloodstream and organ are 8.25 and 2.75, respectively, as predicted by Equations 4.42 and 4.43.

This model illustrates the principle of *partitioning* of a substance between compartments in a closed system. Note that at t equal ∞, the following relations are obtained:

$$(4.42)\quad N_b(\infty) = [\lambda_{ob}N_b(0)/[(\lambda_{bo} + \lambda_{ob})]] + [\lambda_{ob}N_o(0)/(\lambda_{bo} + \lambda_{ob})]$$

$$= [\lambda_{ob}N_b(0) + \lambda_{ob}N_o(0)]/(\lambda_{bo} + \lambda_{ob})]$$

$$= \lambda_{ob}[N_b(0) + N_o(0)]/(\lambda_{bo} + \lambda_{ob})]$$

$$(4.43)\quad N_o(\infty) = [\lambda_{bo}N_b(0)/(\lambda_{bo} + \lambda_{ob})] + [N_o(0)\lambda_{bo}/(\lambda_{bo} + \lambda_{ob})]$$

$$= [\lambda_{bo}N_b(0) + \lambda_{bo}N_o(0)]/(\lambda_{bo} + \lambda_{ob})]$$

$$= \lambda_{bo}[N_b(0) + N_o(0)]/(\lambda_{bo} + \lambda_{ob})]$$

From Equations 4.42 and 4.43 it may be seen that the final amount of the pesticide in each compartment of the closed, coupled system is independent of how the pesticide was distributed originally. After reaching equilibrium, the pesticide will have partitioned between the two compartments, with a fraction $\lambda_{bo}/(\lambda_{bo} + \lambda_{ob})$ in the organ and a fraction $\lambda_{ob}/(\lambda_{bo} + \lambda_{ob})$ in the bloodstream.

4.5.2. Open coupled systems

Figure 4.4 shows another slightly simplified version of Figure 4.1. The system still consists of the bloodstream and an organ with which a pesticide is exchanged. It is an open system, however, because the pesticide moves from the outside (from the GI tract) into the bloodstream at a rate R, and from the organ to some point outside the system. At t equal 0, there is an amount of pesticide equal to $N_b(0)$ in the bloodstream and $N_o(0)$ in the organ. The pesticide then transfers from the blood to the organ with a first-order transfer rate constant λ_{bo}, from the organ back to the blood with a first-order transfer rate constant λ_{ob}, and from the organ to outside the system with a first-order transfer rate constant λ_{out}.

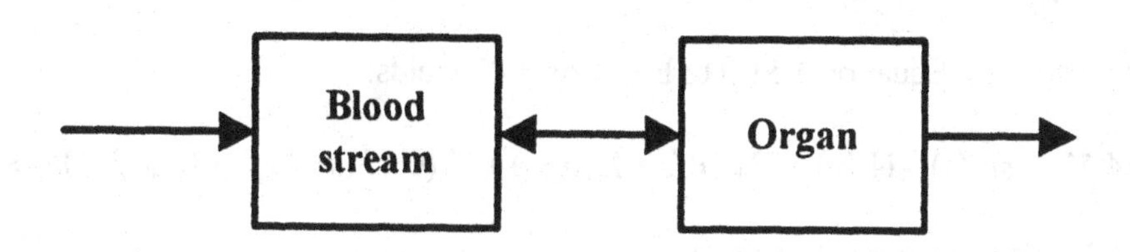

Figure 4.4. A simplified version of Figure 4.1 with bi-directional flow between two compartments but with the substance entering the system from the outside and leaving the system to the outside.

The two differential equations for this system (one for each compartment) are given by:

(4.44) $$dN_b(t)/dt = R + \lambda_{ob}N_o(t) - \lambda_{bo}N_b(t)$$

(4.45) $$dN_o(t)/dt = \lambda_{bo}N_b(t) - \lambda_{ob}N_o(t) - \lambda_{out}N_o(t)$$

Note that this system of equations also cannot be solved using Bernoulli's method since the solution to the first equation would require the solution to the second, and the solution to the second would require the solution to the first.

Applying Theorem 1.3 to both equations, the following relations are obtained:

(4.46) $$\pounds\{dN_b(t)/dt\} = \pounds\{R\} + \pounds\{\lambda_{ob}N_o(t)\} - \pounds\{\lambda_{bo}N_b(t)\}$$

(4.47) $$\pounds\{dN_o(t)/dt\} = \pounds\{\lambda_{bo}N_b(t)\} - \pounds\{(\lambda_{ob}+\lambda_{out})N_o(t)\}$$

Using Theorem 1.7 for the left hand sides and Theorem 1.3 again for the right hand sides yields:

(4.48) $$sn_b(s) - N_b(0) = R/s + \lambda_{ob}n_o(s) - \lambda_{bo}n_b(s)$$

(4.49) $$sn_o(s) - N_o(0) = \lambda_{bo}n_b(s) - (\lambda_{ob}+\lambda_{out})n_o(s)$$

Using Equation 4.48 to solve for $n_b(s)$:

(4.50) $$(s + \lambda_{bo})n_b(s) = R/s + \lambda_{ob}n_o(s) + N_b(0)$$

or

(4.51) $$n_b(s) = [R/s + \lambda_{ob}n_o(s) + N_b(0)] / (s + \lambda_{bo})$$

Substituting Equation 4.51 into Equation 4.49 yields:

(4.52) $$sn_o(s) - N_o(0) = \lambda_{bo}[R/s + \lambda_{ob}n_o(s) + N_b(0)] / (s + \lambda_{bo}) - (\lambda_{ob}+\lambda_{out})n_o(s)$$

Collecting terms involving $n_o(s)$:

(4.53) $$[s + (\lambda_{ob}+\lambda_{out}) - \lambda_{bo}(\lambda_{ob}/(s + \lambda_{bo}))]n_o(s) = \lambda_{bo}R/(s(s + \lambda_{bo})) + \lambda_{bo}N_b(0)/(s + \lambda_{bo}) + N_o(0)$$

Multiplying both sides by $(s+\lambda_{bo})$ yields:

(4.54) $[(s + (\lambda_{ob}+\lambda_{out}))(s + \lambda_{bo}) - \lambda_{bo}\lambda_{ob}]n_o(s) = \lambda_{bo}R/s$

$+ \lambda_{bo}N_b(0) + N_o(0)(s + \lambda_{bo})$

or

(4.55) $n_o(s) = [\lambda_{bo}R/s+\lambda_{bo}N_b(0)+N_o(0)(s + \lambda_{bo})]/[(s+(\lambda_{ob}+\lambda_{out}))(s+\lambda_{bo}) - \lambda_{bo}\lambda_{ob}]$

$N_o(t)$ may then be obtained by first applying the inverse Laplace transform to both sides of the equation:

(4.56) $N_o(t) = \pounds^{-1}\{n_o(s)\} =$

$= \pounds^{-1}\{[\lambda_{bo}R/s+\lambda_{bo}N_b(0)+N_o(0)\ (s + \lambda_{bo})]/[(s+(\lambda_{ob}+\lambda_{out}))(s+\lambda_{bo}) - \lambda_{bo}\lambda_{ob}]\}$

Using Theorem 1.9:

(4.57) $N_o(t) = \pounds^{-1}\{[\lambda_{bo}R]/[s[(s+(\lambda_{ob}+\lambda_{out}))(s+\lambda_{bo}) - \lambda_{bo}\lambda_{ob}]]\}$

$+ \pounds^{-1}\{[\lambda_{bo}N_b(0)]/[(s+(\lambda_{ob}+\lambda_{out}))(s+\lambda_{bo}) - \lambda_{bo}\lambda_{ob}]\}$

$+ \pounds^{-1}\{[N_o(0)(s + \lambda_{bo})\ ]/[(s+(\lambda_{ob}+\lambda_{out}))(s+\lambda_{bo}) - \lambda_{bo}\lambda_{ob}]\}$

The right hand side may be solved using the Heaviside Expansion Theorem applied three times (once to each of the three terms). For the first term:

(4.58) $P(s) = \lambda_{bo}R$

and

(4.59) $Q(s) = s[(s + (\lambda_{ob}+\lambda_{out}))(s + \lambda_{bo}) - \lambda_{bo}\lambda_{ob}]$

$= s[s^2 + s\lambda_{bo} + s(\lambda_{ob}+\lambda_{out}) + (\lambda_{ob}+\lambda_{out})\lambda_{bo} - \lambda_{ob}\lambda_{bo}]$

$= s[s^2 + s(\lambda_{bo} + \lambda_{ob}+\lambda_{out}) + \lambda_{out}\lambda_{bo}]$

$= s^3 + s^2(\lambda_{bo} + \lambda_{ob}+\lambda_{out}) + s\lambda_{out}\lambda_{bo}$

and

(4.60) $$Q^1(s) = 3s^2 + 2s(\lambda_{bo} + \lambda_{ob} + \lambda_{out}) + \lambda_{out}\lambda_{bo}$$

The roots of Equation 4.59 are $\alpha_1 = 0$ and the two roots of $[s^2 + s(\lambda_{bo} + \lambda_{ob} + \lambda_{out}) + \lambda_{out}\lambda_{bo}]$. The latter two roots may be found from the quadratic formula to be:

(4.61) $$\alpha_2 = [-(\lambda_{bo} + \lambda_{ob} + \lambda_{out}) + [(\lambda_{bo} + \lambda_{ob} + \lambda_{out})^2 - 4\lambda_{out}\lambda_{bo}]^{0.5}]/2$$

(4.62) $$\alpha_3 = [-(\lambda_{bo} + \lambda_{ob} + \lambda_{out}) - [(\lambda_{bo} + \lambda_{ob} + \lambda_{out})^2 - 4\lambda_{out}\lambda_{bo}]^{0.5}]/2$$

Using Theorem 1.8:

(4.63) $$\pounds^{-1}\{[\lambda_{bo}R]/[s[(s+(\lambda_{ob}+\lambda_{out}))(s+\lambda_{bo}) - \lambda_{bo}\lambda_{ob}]]\}$$

$$= [\lambda_{bo}R/[3(\alpha_1)^2 + 2(\alpha_1)(\lambda_{bo} + \lambda_{ob} + \lambda_{out}) + \lambda_{out}\lambda_{bo}]]\exp[-\alpha_1 t]$$

$$+ [\lambda_{bo}R/[3(\alpha_2)^2 + 2(\alpha_2)(\lambda_{bo} + \lambda_{ob} + \lambda_{out}) + \lambda_{out}\lambda_{bo}]]\exp[-\alpha_2 t]$$

$$+ [\lambda_{bo}R/[3(\alpha_3)^2 + 2(\alpha_3)(\lambda_{bo} + \lambda_{ob} + \lambda_{out}) + \lambda_{out}\lambda_{bo}]]\exp[-\alpha_3 t]$$

$$= [\lambda_{bo}R/[\lambda_{out}\lambda_{bo}]$$

$$+ [\lambda_{bo}R/[3(\alpha_2)^2 + 2(\alpha_2)(\lambda_{bo} + \lambda_{ob} + \lambda_{out}) + \lambda_{out}\lambda_{bo}]]\exp[-\alpha_2 t]$$

$$+ [\lambda_{bo}R/[3(\alpha_3)^2 + 2(\alpha_3)(\lambda_{bo} + \lambda_{ob} + \lambda_{out}) + \lambda_{out}\lambda_{bo}]]\exp[-\alpha_3 t]$$

For the second term in Equation 4.57:

(4.64) $$P(s) = \lambda_{bo}N_b(0)$$

and

(4.65) $$Q(s) = [(s + (\lambda_{ob}+\lambda_{out}))(s + \lambda_{bo}) - \lambda_{bo}\lambda_{ob}] = s^2 + s\lambda_{bo} + s(\lambda_{ob}+\lambda_{out})$$

$$+ (\lambda_{ob}+\lambda_{out})\lambda_{bo} - \lambda_{ob}\lambda_{bo}$$

$$= s^2 + s(\lambda_{bo}+\lambda_{ob}+\lambda_{out}) + \lambda_{out}\lambda_{bo}$$

and

(4.66) $$Q^1(s) = 2s + (\lambda_{bo}+\lambda_{ob}+\lambda_{out})$$

The roots of Equation 4.65 are the two roots of $s^2 + s(\lambda_{bo}+\lambda_{ob}+\lambda_{out}) + \lambda_{out}\lambda_{bo}$. These two roots may be found from the quadratic formula to be:

(4.67) $\alpha_2 = [-(\lambda_{bo} + \lambda_{ob}+\lambda_{out}) + [(\lambda_{bo} + \lambda_{ob}+\lambda_{out})^2 - 4\lambda_{out}\lambda_{bo}]^{0.5}]/2$

(4.68) $\alpha_3 = [-(\lambda_{bo} + \lambda_{ob}+\lambda_{out}) - [(\lambda_{bo} + \lambda_{ob}+\lambda_{out})^2 - 4\lambda_{out}\lambda_{bo}]^{0.5}]/2$

Note they are the same as the quadratic roots for the first term. Therefore:

(4.69) $£^{-1}\{[\lambda_{bo}N_b(0)]/[(s+(\lambda_{ob}+\lambda_{out}))(s+\lambda_{bo}) - \lambda_{bo}\lambda_{ob}]\}$

$$= [\lambda_{bo}N_b(0)/[2(\alpha_2) + (\lambda_{bo}+\lambda_{ob}+\lambda_{out})]]\exp[-\alpha_2 t]$$

$$+ [\lambda_{bo}N_b(0)/[2(\alpha_3) + (\lambda_{bo}+\lambda_{ob}+\lambda_{out})]]\exp[-\alpha_3 t]$$

For the third term in Equation 4.57:

(4.70) $$P(s) = N_o(0)(s + \lambda_{bo})$$

and

(4.71) $Q(s) = [(s + (\lambda_{ob}+\lambda_{out}))(s + \lambda_{bo}) - \lambda_{bo}\lambda_{ob}] = s^2 + s\lambda_{bo} + s(\lambda_{ob}+\lambda_{out})$

$$+ (\lambda_{ob}+\lambda_{out})\lambda_{bo} - \lambda_{ob}\lambda_{bo}$$

$$= s^2 + s(\lambda_{bo}+\lambda_{ob}+\lambda_{out}) + \lambda_{out}\lambda_{bo}$$

and

(4.72) $$Q^1(s) = 2s + (\lambda_{bo}+\lambda_{ob}+\lambda_{out})$$

The roots of Equation 4.71 are the two roots of $s^2 + s(\lambda_{bo}+\lambda_{ob}+\lambda_{out}) + \lambda_{out}\lambda_{bo}$. These two roots may be found from the quadratic formula to be:

(4.73) $\alpha_2 = [-(\lambda_{bo} + \lambda_{ob}+\lambda_{out}) + [(\lambda_{bo} + \lambda_{ob}+\lambda_{out})^2 - 4\lambda_{out}\lambda_{bo}]^{0.5}]/2$

(4.74) $\alpha_3 = [-(\lambda_{bo} + \lambda_{ob}+\lambda_{out}) - [(\lambda_{bo} + \lambda_{ob}+\lambda_{out})^2 - 4\lambda_{out}\lambda_{bo}]^{0.5}]/2$

Note they are the same as the roots for the second term. Therefore:

(4.75) $\pounds^{-1}\{[N_o(0)(s+\lambda_{bo})]/[(s+(\lambda_{ob}+\lambda_{out}))(s+\lambda_{bo}) - \lambda_{bo}\lambda_{ob}]\}$

$= [N_o(0)(\alpha_2+\lambda_{bo})/[2(\alpha_2)+(\lambda_{bo}+\lambda_{ob}+\lambda_{out})]]\exp[-\alpha_2 t]$

$+ [N_o(0)(\alpha_3+\lambda_{bo})]/[2(\alpha_3)+(\lambda_{bo}+\lambda_{ob}+\lambda_{out})]]\exp[-\alpha_3 t]$

Combining Equations 4.63, 4.69 and 4.75:

(4.76) $N_o(t) = [\lambda_{bo}R]/[\lambda_{out}\lambda_{bo}]$

$+ [\lambda_{bo}R/[3(\alpha_2)^2+2(\alpha_2)(\lambda_{bo}+\lambda_{ob}+\lambda_{out})+\lambda_{out}\lambda_{bo}]]\exp[-\alpha_2 t]$

$+ [\lambda_{bo}R/[3(\alpha_3)^2+2(\alpha_3)(\lambda_{bo}+\lambda_{ob}+\lambda_{out})+\lambda_{out}\lambda_{bo}]]\exp[-\alpha_2 t]$

$+ [\lambda_{bo}N_b(0)/[2(\alpha_2)+(\lambda_{bo}+\lambda_{ob}+\lambda_{out})]]\exp[-\alpha_2 t]$

$+ [\lambda_{bo}N_b(0)/[2(\alpha_3)+(\lambda_{bo}+\lambda_{ob}+\lambda_{out})]]\exp[-\alpha_3 t]$

$+ [N_o(0)\,(\alpha_2+\lambda_{bo})/[2(\alpha_2)+(\lambda_{bo}+\lambda_{ob}+\lambda_{out})]]\exp[-\alpha_2 t]$

$+ [N_o(0)\,(\alpha_2+\lambda_{bo})/[2(\alpha_3)+(\lambda_{bo}+\lambda_{ob}+\lambda_{out})]]\exp[-\alpha_3 t]$

As in Section 4.5.2, there are now 2 ways to solve for $N_b(t)$. The solution for $N_o(t)$ might be substituted back into Equation 4.44. This would yield a differential equation for $N_b(t)$ involving only $N_b(t)$. Bernoulli's method might then be applied. An alternative is to substitute Equation 4.55 into Equation 4.48 and continue using Laplace transform methods. The latter approach provides additional practice at the use of Laplace transforms, and so is adopted here. Making the indicated substitution:

(4.77) $sn_b(s) - N_b(0) =$

$R/s + \lambda_{ob}[\lambda_{bo}R/s+\lambda_{bo}N_b(0)+N_o(0)(s+\lambda_{bo})]/[(s+(\lambda_{ob}+\lambda_{out}))(s+\lambda_{bo})-\lambda_{bo}\lambda_{ob}] - \lambda_{bo}n_b(s)$

Collecting terms involving $n_b(s)$:

(4.78) $(s+\lambda_{bo})n_b(s) =$

$N_b(0) + R/s + \lambda_{ob}[\lambda_{bo}R/s+\lambda_{bo}N_b(0)+N_o(0)(s+\lambda_{bo})]/[(s+(\lambda_{ob}+\lambda_{out}))(s+\lambda_{bo})-\lambda_{bo}\lambda_{ob}]$

Dividing both sides by $(s + \lambda_{bo})$ yields:

(4.79) $$n_b(s) = N_b(0)/(s+\lambda_{bo}) + R/[s(s+\lambda_{bo})]$$

$$+ \lambda_{ob}\lambda_{bo}R/[s(s+\lambda_{bo})[(s+(\lambda_{ob}+\lambda_{out}))(s+\lambda_{bo})-\lambda_{bo}\lambda_{ob}]]$$

$$+ \lambda_{ob}\lambda_{bo}N_b(0)/[(s+\lambda_{bo})[(s+(\lambda_{ob}+\lambda_{out}))(s+\lambda_{bo})-\lambda_{bo}\lambda_{ob}]]$$

$$+ \lambda_{ob}N_o(0)/[(s+(\lambda_{ob}+\lambda_{out}))(s+\lambda_{bo})-\lambda_{bo}\lambda_{ob}]$$

This solution involves five terms. Applying the inverse Laplace transform to all terms using Theorem 1.9:

(4.80) $$N_b(t) = £^{-1}\{n_b(s)\} =$$

$$£^{-1}\{N_b(0)/(s+\lambda_{bo})\} + £^{-1}\{R/[s(s+\lambda_{bo})]\}$$

$$+ £^{-1}\{\lambda_{ob}\lambda_{bo}R/[s(s+\lambda_{bo})[(s+(\lambda_{ob}+\lambda_{out}))(s+\lambda_{bo})-\lambda_{bo}\lambda_{ob}]]\}$$

$$+ £^{-1}\{\lambda_{ob}\lambda_{bo}N_b(0)/[(s+\lambda_{bo})[(s+(\lambda_{ob}+\lambda_{out}))(s+\lambda_{bo})-\lambda_{bo}\lambda_{ob}]]\}$$

$$+ £^{-1}\{\lambda_{ob}N_o(0)/[(s+(\lambda_{ob}+\lambda_{out}))(s+\lambda_{bo})-\lambda_{bo}\lambda_{ob}]\}$$

For the first term on the right hand side of Equation 4.80:

(4.81) $$P(s) = N_b(0)$$

and

(4.82) $$Q(s) = (s + \lambda_{bo})$$

and

(4.83) $$Q^1(s) = 1$$

The roots of Equation 4.82 are $\alpha_1 = -\lambda_{bo}$. Using Theorem 1.8:

(4.84) $$£^{-1}\{N_b(0)/(s+\lambda_{bo})\} = [N_b(0)/1]\exp[-\lambda_{bo}t] = N_b(0)\exp[-\lambda_{bo}t]$$

For the second term on the right hand side of Equation 4.80:

(4.85) $$P(s) = R$$

and

(4.86) $$Q(s) = s(s + \lambda_{bo}) = (s^2 + s\lambda_{bo})$$

and

(4.87) $$Q^1(s) = 2s + \lambda_{bo}$$

The roots of Equation 4.86 are $\alpha_1 = 0$ and $\alpha_2 = -\lambda_{bo}$. Using Theorem 1.8:

(4.88) $$£^{-1}\{R/[s(s+\lambda_{bo})]\} =$$

$$= [R/(2\bullet 0 + \lambda_{bo})]\exp[-0t] + [R/(2\bullet(-\lambda_{bo}) + \lambda_{bo})]\exp[-\lambda_{bo}t]$$

$$= [R/\lambda_{bo}] - [R/\lambda_{bo}]\exp[-\lambda_{bo}t] = [R/\lambda_{bo}](1 - \exp[-\lambda_{bo}t])$$

For the third term on the right hand side of Equation 4.80:

(4.89) $$P(s) = \lambda_{ob}\lambda_{bo}R$$

and

(4.90) $$Q(s) = s(s+\lambda_{bo})[(s + (\lambda_{ob}+\lambda_{out}))(s + \lambda_{bo}) - \lambda_{bo}\lambda_{ob}]$$

$$= s(s+\lambda_{bo})[s^2 + s\lambda_{bo} + s(\lambda_{ob}+\lambda_{out}) + (\lambda_{ob}+\lambda_{out})\lambda_{bo} - \lambda_{ob}\lambda_{bo}]$$

$$= s(s+\lambda_{bo})[s^2 + s(\lambda_{bo} + \lambda_{ob}+\lambda_{out}) + \lambda_{out}\lambda_{bo}]$$

$$= s(s^3 + s^2(\lambda_{bo} + \lambda_{ob}+\lambda_{out}) + s\lambda_{out}\lambda_{bo}+\lambda_{bo}s^2 + \lambda_{bo}s(\lambda_{bo} + \lambda_{ob}+\lambda_{out}) + \lambda_{bo}\lambda_{out}\lambda_{bo})$$

$$= s^4 + s^3(2\lambda_{bo} + \lambda_{ob}+\lambda_{out}) + s^2(2\lambda_{out}\lambda_{bo}+ \lambda_{bo}\lambda_{bo} + \lambda_{bo}\lambda_{ob}) + s\lambda_{bo}\lambda_{out}\lambda_{bo}$$

and

(4.91) $$Q^1(s) = 4s^3 + 3s^2(2\lambda_{bo} + \lambda_{ob}+\lambda_{out}) + 2s(2\lambda_{out}\lambda_{bo}+ \lambda_{bo}\lambda_{bo} + \lambda_{bo}\lambda_{ob}) + \lambda_{bo}\lambda_{out}\lambda_{bo}$$

The roots of Equation 4.90 are $\alpha_1 = -\lambda_{bo}$, $\alpha_4 = 0$ and the two roots of $s^2 + s(\lambda_{bo}+\lambda_{ob}+\lambda_{out}) + \lambda_{out}\lambda_{bo}$. These two roots may be found from the quadratic formula to be:

(4.92) $\alpha_2 = [-(\lambda_{bo} + \lambda_{ob}+\lambda_{out}) + [(\lambda_{bo} + \lambda_{ob}+\lambda_{out})^2 - 4\lambda_{out}\lambda_{bo}]^{0.5}]/2$

(4.93) $\alpha_3 = [-(\lambda_{bo} + \lambda_{ob}+\lambda_{out}) - [(\lambda_{bo} + \lambda_{ob}+\lambda_{out})^2 - 4\lambda_{out}\lambda_{bo}]^{0.5}]/2$

Therefore:

(4.94) $£^{-1}\{\lambda_{ob}\lambda_{bo}R/[s(s+\lambda_{bo})[(s+(\lambda_{ob}+\lambda_{out}))(s+\lambda_{bo})-\lambda_{bo}\lambda_{ob}]]\}$

$$= [\lambda_{ob}\lambda_{bo}R/[4(0)^3 + 3(0)^2(2\lambda_{bo} + \lambda_{ob}+\lambda_{out}) + 2(0)(2\lambda_{out}\lambda_{bo}+ \lambda_{bo}\lambda_{bo} + \lambda_{bo}\lambda_{ob}) + \lambda_{bo}\lambda_{out}\lambda_{bo}]]\exp[0t]$$

$$+ [\lambda_{ob}\lambda_{bo}R/[4(-\lambda_{bo})^3 + 3(-\lambda_{bo})^2(2\lambda_{bo} + \lambda_{ob}+\lambda_{out}) + + 2(-\lambda_{bo})(2\lambda_{out}\lambda_{bo}+ \lambda_{bo}\lambda_{bo} + \lambda_{bo}\lambda_{ob}) + \lambda_{bo}\lambda_{out}\lambda_{bo}]]\exp[-\lambda_{bo}t]$$

$$+ [\lambda_{ob}\lambda_{bo}R/[4(\alpha_2)^3 + 3(\alpha_2)^2(2\lambda_{bo} + \lambda_{ob}+\lambda_{out}) + + 2(\alpha_2)(2\lambda_{out}\lambda_{bo}+ \lambda_{bo}\lambda_{bo} + \lambda_{bo}\lambda_{ob}) + \lambda_{bo}\lambda_{out}\lambda_{bo}]]\exp[(\alpha_2)t]$$

$$+ [\lambda_{ob}\lambda_{bo}R/[4(\alpha_3)^3 + 3(\alpha_3)^2(2\lambda_{bo} + \lambda_{ob}+\lambda_{out}) + + 2(\alpha_3)(2\lambda_{out}\lambda_{bo}+ \lambda_{bo}\lambda_{bo} + \lambda_{bo}\lambda_{ob}) + \lambda_{bo}\lambda_{out}\lambda_{bo}]]\exp[-\lambda_{bo}t]$$

$$= \lambda_{ob}\lambda_{bo}R/[\lambda_{bo}\lambda_{out}\lambda_{bo}]$$

$$+ [\lambda_{ob}\lambda_{bo}R/[4(-\lambda_{bo})^3 + 3(-\lambda_{bo})^2(2\lambda_{bo} + \lambda_{ob}+\lambda_{out}) + + 2(-\lambda_{bo})(2\lambda_{out}\lambda_{bo}+ \lambda_{bo}\lambda_{bo} + \lambda_{bo}\lambda_{ob}) + \lambda_{bo}\lambda_{out}\lambda_{bo}]]\exp[-\lambda_{bo}t]$$

$$+ [\lambda_{ob}\lambda_{bo}R/[4(\alpha_2)^3 + 3(\alpha_2)^2(2\lambda_{bo} + \lambda_{ob}+\lambda_{out}) + + 2(\alpha_2)(2\lambda_{out}\lambda_{bo}+ \lambda_{bo}\lambda_{bo} + \lambda_{bo}\lambda_{ob}) + \lambda_{bo}\lambda_{out}\lambda_{bo}]]\exp[(\alpha_2)t]$$

$$+ [\lambda_{ob}\lambda_{bo}R/[4(\alpha_3)^3 + 3(\alpha_3)^2(2\lambda_{bo} + \lambda_{ob}+\lambda_{out}) + + 2(\alpha_3)(2\lambda_{out}\lambda_{bo}+ \lambda_{bo}\lambda_{bo} + \lambda_{bo}\lambda_{ob}) + \lambda_{bo}\lambda_{out}\lambda_{bo}]]\exp[-\lambda_{bo}t]$$

For the fourth term on the right hand side of Equation 4.80:

(4.95) $$P(s) = \lambda_{ob}\lambda_{bo}N_b(0)$$

and

(4.96) $$Q(s) = (s+\lambda_{bo})[(s + (\lambda_{ob}+\lambda_{out}))(s + \lambda_{bo}) - \lambda_{bo}\lambda_{ob}]$$

$$= (s+\lambda_{bo})[s^2 + s\lambda_{bo} + s(\lambda_{ob}+\lambda_{out}) + (\lambda_{ob}+\lambda_{out})\lambda_{bo} - \lambda_{ob}\lambda_{bo}]$$

$$= (s+\lambda_{bo})[s^2 + s(\lambda_{bo} + \lambda_{ob}+\lambda_{out}) + \lambda_{out}\lambda_{bo}]$$

$$= s^3 + s^2(\lambda_{bo} + \lambda_{ob}+\lambda_{out}) + s\lambda_{out}\lambda_{bo}+\lambda_{bo}s^2 + \lambda_{bo}s(\lambda_{bo} + \lambda_{ob}+\lambda_{out}) + \lambda_{bo}\lambda_{out}\lambda_{bo}$$

$$= s^3 + s^2(2\lambda_{bo} + \lambda_{ob}+\lambda_{out}) + s(2\lambda_{out}\lambda_{bo}+ \lambda_{bo}\lambda_{bo} + \lambda_{bo}\lambda_{ob}) + \lambda_{bo}\lambda_{out}\lambda_{bo}$$

and

(4.97) $$Q^1(s) = 3s^2 + 2s(2\lambda_{bo} + \lambda_{ob}+\lambda_{out}) + (2\lambda_{out}\lambda_{bo}+ \lambda_{bo}\lambda_{bo} + \lambda_{bo}\lambda_{ob})$$

The roots of Equation 4.96 are $\alpha_1 = -\lambda_{bo}$ and the two roots of $s^2 + s(\lambda_{bo}+\lambda_{ob}+\lambda_{out}) + \lambda_{out}\lambda_{bo}$. These two roots may be found from the quadratic formula to be:

(4.98) $$\alpha_2 = [-(\lambda_{bo} + \lambda_{ob}+\lambda_{out}) + [(\lambda_{bo} + \lambda_{ob}+\lambda_{out})^2 - 4\lambda_{out}\lambda_{bo}]^{0.5}]/2$$

(4.99) $$\alpha_3 = [-(\lambda_{bo} + \lambda_{ob}+\lambda_{out}) - [(\lambda_{bo} + \lambda_{ob}+\lambda_{out})^2 - 4\lambda_{out}\lambda_{bo}]^{0.5}]/2$$

Therefore:

(4.100) $$£^{-1}\{\lambda_{ob}\lambda_{bo}N_b(0)/[(s+\lambda_{bo})[(s+(\lambda_{ob}+\lambda_{out}))(s+\lambda_{bo})-\lambda_{bo}\lambda_{ob}]]\}$$

$$= [\lambda_{ob}\lambda_{bo}N_b(0)/[3(-\lambda_{bo})^2+2(-\lambda_{bo})(2\lambda_{bo}+\lambda_{ob}+\lambda_{out})$$

$$+(2\lambda_{out}\lambda_{bo}+\lambda_{bo}\lambda_{bo}+\lambda_{bo}\lambda_{ob})]]\exp[-\lambda_{bo}t]$$

$$+ [\lambda_{ob}\lambda_{bo}N_b(0)/[3(\alpha_2)^2+2(\alpha_2)(2\lambda_{bo}+\lambda_{ob}+\lambda_{out})$$

$$+(2\lambda_{out}\lambda_{bo}+\lambda_{bo}\lambda_{bo}+\lambda_{bo}\lambda_{ob})]]\exp[\alpha_2 t]$$

$$+ [\lambda_{ob}\lambda_{bo}N_b(0)/[3(\alpha_3)^2+2(\alpha_3)(2\lambda_{bo}+\lambda_{ob}+\lambda_{out})$$

$$+(2\lambda_{out}\lambda_{bo}+\lambda_{bo}\lambda_{bo}+\lambda_{bo}\lambda_{ob})]]\exp[\alpha_3 t]$$

For the fifth term on the right hand side of Equation 4.80:

$$(4.101) \qquad P(s) = \lambda_{ob}N_o(0)$$

and

$$(4.102) \quad Q(s) = [(s+(\lambda_{ob}+\lambda_{out}))(s+\lambda_{bo}) - \lambda_{bo}\lambda_{ob}] =$$

$$= s^2 + s\lambda_{bo} + s(\lambda_{ob}+\lambda_{out}) + (\lambda_{ob}+\lambda_{out})\lambda_{bo} - \lambda_{ob}\lambda_{bo}$$

$$= s^2 + s(\lambda_{bo}+\lambda_{ob}+\lambda_{out}) + \lambda_{out}\lambda_{bo}$$

and

$$(4.103) \qquad Q^1(s) = 2s + (\lambda_{bo}+\lambda_{ob}+\lambda_{out})$$

The roots of Equation 4.102 are the two roots of $s^2 + s(\lambda_{bo}+\lambda_{ob}+\lambda_{out}) + \lambda_{out}\lambda_{bo}$. These two roots may be found from the quadratic formula to be:

$$(4.104) \quad \alpha_2 = [-(\lambda_{bo} + \lambda_{ob}+\lambda_{out}) + [(\lambda_{bo} + \lambda_{ob}+\lambda_{out})^2 - 4\lambda_{out}\lambda_{bo}]^{0.5}]/2$$

$$(4.105) \quad \alpha_3 = [-(\lambda_{bo} + \lambda_{ob}+\lambda_{out}) - [(\lambda_{bo} + \lambda_{ob}+\lambda_{out})^2 - 4\lambda_{out}\lambda_{bo}]^{0.5}]/2$$

Therefore:

$$(4.106) \quad \pounds^{-1}\{[\lambda_{ob}N_o(0)]/[(s+(\lambda_{ob}+\lambda_{out}))(s+\lambda_{bo}) - \lambda_{bo}\lambda_{ob}]\}$$

$$= [\lambda_{ob}N_o(0)/[2(\alpha_2) + (\lambda_{bo}+\lambda_{ob}+\lambda_{out})]]\exp[-\alpha_2 t]$$

$$+ [\lambda_{ob}N_o(0)/[2(\alpha_3) + (\lambda_{bo}+\lambda_{ob}+\lambda_{out})]]\exp[-\alpha_3 t]$$

Combining Equations 4.84, 4.88, 4.94, 4.100 and 4.106:

$$(4.107) \quad N_b(t) = N_b(0)\exp[-\lambda_{bo}t] + [R/\lambda_{bo}](1 - \exp[-\lambda_{bo}t])$$

$$+ \lambda_{ob}\lambda_{bo}R/[\lambda_{bo}\lambda_{out}\lambda_{bo}]$$

$$+ [\lambda_{ob}\lambda_{bo}R/[4(-\lambda_{bo})^3 + 3(-\lambda_{bo})^2(2\lambda_{bo} + \lambda_{ob}+\lambda_{out})$$

$$+ 2(-\lambda_{bo})(2\lambda_{out}\lambda_{bo} + \lambda_{bo}\lambda_{bo} + \lambda_{bo}\lambda_{ob}) + \lambda_{bo}\lambda_{out}\lambda_{bo}]]\exp[-\lambda_{bo}t]$$

$$+ [\lambda_{ob}\lambda_{bo}R/[4(\alpha_2)^3 + 3(\alpha_2)^2(2\lambda_{bo} + \lambda_{ob}+\lambda_{out}) +$$

$$+ 2(\alpha_2)\,(2\lambda_{out}\lambda_{bo} + \lambda_{bo}\lambda_{bo} + \lambda_{bo}\lambda_{ob}) + \lambda_{bo}\lambda_{out}\lambda_{bo}]]\exp[(\alpha_2)t]$$

$$+ [\lambda_{ob}\lambda_{bo}R/[4(\alpha_3)^3 + 3(\alpha_3)^2(2\lambda_{bo} + \lambda_{ob}+\lambda_{out}) +$$

$$+ 2(\alpha_3)(2\lambda_{out}\lambda_{bo} + \lambda_{bo}\lambda_{bo} + \lambda_{bo}\lambda_{ob}) + \lambda_{bo}\lambda_{out}\lambda_{bo}]]\exp[-\lambda_{bo}t]$$

$$+ [\lambda_{ob}\lambda_{bo}N_b(0)/[3(-\lambda_{bo})^2+2(-\lambda_{bo})(2\lambda_{bo}+\lambda_{ob}+\lambda_{out})$$

$$+(2\lambda_{out}\lambda_{bo}+\lambda_{bo}\lambda_{bo}+\lambda_{bo}\lambda_{ob})]]\exp[-\lambda_{bo}t]$$

$$+ [\lambda_{ob}\lambda_{bo}N_b(0)/[3(\alpha_2)^2+2(\alpha_2)(2\lambda_{bo}+\lambda_{ob}+\lambda_{out})$$

$$+(2\lambda_{out}\lambda_{bo}+\lambda_{bo}\lambda_{bo}+\lambda_{bo}\lambda_{ob})]]\exp[\alpha_2 t]$$

$$+ [\lambda_{ob}\lambda_{bo}N_b(0)/[3(\alpha_3)^2+2(\alpha_3)(2\lambda_{bo}+\lambda_{ob}+\lambda_{out})$$

$$+(2\lambda_{out}\lambda_{bo}+\lambda_{bo}\lambda_{bo}+\lambda_{bo}\lambda_{ob})]]\exp[\alpha_3 t]$$

$$+ [\lambda_{ob}N_o(0)/[2(\alpha_2) + (\lambda_{bo}+\lambda_{ob}+\lambda_{out})]]\exp[-\alpha_2 t]$$

$$+ [\lambda_{ob}N_o(0)/[2(\alpha_3) + (\lambda_{bo}+\lambda_{ob}+\lambda_{out})]]\exp[-\alpha_3 t]$$

Note the complexity, or at least the length, of analytic solutions for N(t) in even the simplest open, coupled systems. For larger systems (i.e. with a larger number of coupled compartments), analytic solutions either are extremely difficult or even impossible to obtain. In these cases, numerical solutions are necessary, as discussed in Chapter 6.

4.6. Some Additional Laplace Transforms

This section presents a number of Laplace transforms that may be encountered in solving coupled systems of differential equations, or differential equations of higher than first order. In each case, the function N(t) is shown first, followed by its corresponding inverse Laplace transform, n(s). In looking for the Laplace transform of a function, locate the equivalent N(t) and find the corresponding n(s). In looking for the inverse Laplace transform, locate the equivalent n(s) and find the corresponding N(t). In other words, the following entries may be used for both the transform and inverse transform steps of a solution.

N(t)	n(s)
t	$1 / s^2$
e^{at}	$1 / (s-a)$
$\cos(at)$	$s / (s^2+a^2)$
$t \cos(at)$	$(s^2-a^2) / (s^2+a^2)^2$
$\cosh(at)$	$s / (s^2-a^2)$
$\sin(at) / a$	$1 / (s^2+a^2)$
$\sinh(at) / a$	$1 / (s^2-a^2)$
$(e^{bt} - e^{at}) / (b-a)$	$1 / ((s-a)(s-b))$
$(be^{bt} - ae^{at}) / (b-a)$	$s / ((s-a)(s-b))$
$t \sinh(at) / 2a$	$s / (s^2-a^2)^2$
$\mathrm{Ln}(t)$	$-(0.5772 + \mathrm{Ln}(s)) / s$

References

1. W. Schlesinger, *Biogeochemistry: An Analysis of Global Change*, Academic Press, New York, 1997.
2. A. Ford, *Modeling the Environment: An Introduction to System Dynamics Modeling of Environmental Systems*, Island Press, Washington, DC, 1999.
3. M. Allaby, *Basics of Environmental Science*, Routledge, London, 1996.
4. M Spiegel, *Laplace Transforms*, Schaum Publishing Co., NY, 1965
5. D. Crawford-Brown, *Theoretical and Mathematical Foundations of Human Health Risk Analysis*, Kluwer Academic Publishers, 1997.
6. F. Nixon, *Handbook of Laplace Transformation*, Prentice-Hall, Englewood Cliffs, NJ, 1960.

CHAPTER 5
Matrix Methods and Spectral Analysis

5.1. Spectra in Environmental Problems

Environmental systems may in some cases be described on the basis of the output, or *response*, of the system to an input, or *perturbation* [1]. Even if the internal structure of the system is not understood, such input-output analysis can provide at least a partial basis for predicting the evolution of the system. For example, an ecosystem might consist of a number, N, of unique species. Each species might have some sensitivity to each of several pollutants present in the environment, meaning the species will respond to the presence of that pollutant (i.e. the pollutant perturbs the species, with the response being a change in the state vector describing the health of that species). If each species is exposed to the full range of pollutants simultaneously, the total response of each species is related to the composite effect of all of the pollutants acting simultaneously.

It is possible at times to determine the state of the environment causing the changes in the state of health by examining the responses in several species exposed to the same environment. Such a process is one of *unfolding*, since the environmental state is unfolded from the response of the ecosystem. Similar problems of unfolding arise in the interpretation of instrument response to a spectrum of environmental insults (e.g. the response of a radiation detector to a spectrum of radiation [2]), and so the methods discussed in this chapter fall into the general class of *spectral unfolding*. In fact, the example in this chapter treats an ecosystem as an instrument, and the environment as a spectrum of perturbations to that instrument. The task is to determine the spectrum of the perturbation from information on the response of the instrument.

Imagine an ecosystem consisting of two different species. There are two forms of pollution in the environment, with concentrations of C_1 and C_2. Both species show some effect (such as death) in the presence of each of the two pollutants, with the probability of the effect proportional to the concentration of the pollutant in the environment. The probability of the effect is shown as R_i, where R stands for *response* and the subscript i indicates the ith species in the ecosystem. This probability is described by a *sensitivity*, S, which is the proportionality constant between the concentration of the pollutant in the environment and the probability of effect. In general, this sensitivity will be shown as S_{ij}, where subscript i indicates the species and j indicates the pollutant. The units of S_{ij} are probability of response (or, at times, severity of resonse) per unit concentration in the environment; e,g. $(g/m^3)^{-1}$. There then are two equations defining the response of this system:

$$R_1 = S_{1,1}C_1 + S_{1,2}C_2 \tag{5.1}$$

$$R_2 = S_{2,1}C_1 + S_{2,2}C_2 \tag{5.2}$$

Imagine now that both the sensitivity and the response is known, but that the concentration is not known. The problem shown in Equations 5.1 and 5.2 can still be solved by noting that there are 2 equations with two unknowns (the two concentrations C_1 and C_2).

Solutions to these equations may be found by simple *Gaussian back-elimination* as discussed in the next section. When the system becomes complex, however, with many species present, back-elimination is time consuming and inefficient. In such cases, methods of matrix analysis are more suitable. We might rewrite Equations 5.1 and 5.2 as:

$$[R] = [S][C] \tag{5.3}$$

where [R], [S] and [C] are all *matrices*. Since [R] and [S] both are known, it will be possible to solve for [C] using the methods of *matrix analysis* provided in this chapter [3]. Such methods provide a computationally efficient way to unfold perturbation spectra from response and sensitivity spectra. They are particularly useful when the numbers of species and environmental insults are so large that back-elimination becomes infeasible.

5.2. Back-elimination

The simplest method (but not necessarily the most efficient) for solving systems of algebraic equations such as Equations 5.1 and 5.2 is back-elimination [4]. In fact, essentially all of the matrix methods presented later in this chapter accomplish the same tasks as back-elimination, although the conceptual links between the methods may not be self-evident.

Back-elimination involves use of one equation to solve for one unknown in terms of the other unknowns, substitution of this solution into one of the other equations in the series, and a continuation of the process until all unknowns are expressed solely in terms of known quantities. Considering Equation 5.1, one could solve for either of the unknowns (c_1 or c_2); it does not matter which is used first, the answer will be the same in the end. Note that in the equations below, the terms have been shown in lower case rather than the upper case form of Equations 5.1 and 5.2. This is a common convention when dealing with matrix problems and will be continued throughout this chaper.

Beginning with Equation 5.1 and solving for c_1 in terms of c_2:

(5.4) $$c_1 = (r_1 - s_{1,2}c_2)/s_{1,1}$$

Substituting Equation 5.4 into Equation 5.2:

(5.5) $$r_2 = s_{2,1}c_1 + s_{2,2}c_2 = (s_{2,1}(r_1 - s_{1,2}c_2)/s_{1,1}) + s_{2,2}c_2$$

Solving for c_2:

(5.6) $$r_2 = (s_{2,1}r_1/s_{1,1}) - (s_{2,1}s_{1,2}c_2/s_{1,1}) + (s_{2,2}c_2)$$

$$= (s_{2,1}r_1/s_{1,1}) - ((s_{2,1}s_{1,2}/s_{1,1}) - s_{2,2})c_2$$

or

(5.7) $$((s_{2,1}s_{1,2}/s_{1,1}) - s_{2,2})c_2 = (s_{2,1}r_1/s_{1,1}) - r_2$$

or

(5.8) $$c_2 = ((s_{2,1}r_1/s_{1,1}) - r_2) / ((s_{2,1}s_{1,2}/s_{1,1}) - s_{2,2})$$

Note that c_2 is now entirely a function of several known quantities, and so its numerical value may be calculated through substitution of these numerical values into Equation 5.8. Substituting Equation 5.8 into Equation 5.1:

(5.9) $$r_1 = s_{1,1}c_1 + s_{1,2}((s_{2,1}r_1/s_{1,1}) - r_2) / ((s_{2,1}s_{1,2}/s_{1,1}) - s_{2,2})$$

or

(5.10) $$c_1 = (r_1 - (s_{1,2}((s_{2,1}r_1/s_{1,1}) - r_2) / ((s_{2,1}s_{1,2}/s_{1,1}) - s_{2,2})))/s_{1,1}$$

Back-elimination, also known as the *method of substitution*, generally will result in a unique solution; i.e. the solution that follows from the method will be the only solution satisfying the system of equations. An exception arises, however, when the equations are linearly dependent, or multiples of one another. If, for example, $s_{1,1} = bs_{2,1}$ and $s_{1,2} = bs_{2,2}$ in Equations 5.1 and 5.2, where b is any constant, then Equation 5.2 is simply a multiple (or proportional to) Equation 5.1. Equation 5.2 is *linearly dependent* on Equation 5.1. This leads to the following equations:

(5.1b) $$r_1 = s_{1,1}c_1 + s_{1,2}c_2$$

(5.2b) $$r_2 = s_{2,1}c_1 + s_{2,2}c_2$$

Making the substitutions for the sensitivities noted above:

(5.1c) $$r_1 = bs_{2,1}c_1 + bs_{2,2}c_2$$

(5.2c) $$r_2 = s_{2,1}c_1 + s_{2,2}c_2$$

Solving for c2 using Equation 5.2c:

(5.2d) $$c_2 = (r_2 - s_{2,1}c_1) / s_{2,2}$$

Substituting c_2 back into Equation 5.1c:

(5.1d) $$r_1 = bs_{2,1}c_1 + bs_{2,2} (r_2 - s_{2,1}c_1) / s_{2,2}$$

or

(5.1e) $$r_1 = bs_{2,1}c_1 + b (r_2 - s_{2,1}c_1)$$

or

(5.1f) $$r_1 = b\, r_2$$

Note that Equation 5.1f does not involve either of the concentrations. In such cases, there are no unique solutions but rather an infinite number of combinations of c_1 and c_2, an infinite number of sets (c_1,c_2), satisfying the set of equations.

5.3. Matrices

Back-elimination may always be used in solving systems of equations containing N equations and N unknowns. It becomes unwieldy, however, when N becomes large, as is often the case in environmental phenomena. For large values of N, it is easier computationally to treat the problem as one of matrix analysis, and so we turn to those methods of analysis in the rest of this chapter. We will continue with the example of Equation 3, in which [R] = [S][C].

Example 5.1. For Equations 5.1 and 5.2, let r_1 and r_2 be 0.7 and 1, respectively. Let $s_{1,1}$ and $s_{1,2}$ be 0.1 and 0.3, respectively. Let $s_{2,1}$ and $s_{2,2}$ be 0.2 and 0.4, respectively. This yields the equations:

$$0.7 = 0.1c_1 + 0.3c_2 \quad \text{and} \quad 1 = 0.2c_1 + 0.4c_2$$

From Equation 5.8, $c_2 = ((0.2\bullet 0.7/0.1)-1) / ((0.2\bullet 0.3/0.1) - 0.4) = 2$
From Equation 5.10, $c_1 = (0.7-(0.3\,((0.2\bullet 0.7/0.1)-1)/((0.2\bullet 0.3/0.1)-0.4)))/0.1 = 1$

Checking the solutions using Equations 5.1 and 5.2:

$0.7 = 0.1\bullet 1 + 0.3\bullet 2$

$1 = 0.2\bullet 1 + 0.4\bullet 2$

So the original equations are satisfied by these solutions.

A matrix consists of a set of numbers organized into *rows* (horizontal on this page) and *columns* (vertical on this page). Each of the numbers on the array is an *element* of the matrix. If the matrix is M (often shown as [M] or |M|, but we will show it simply as M in this book), by convention each element is designated $m_{i,j}$. The subscript i indicates the row number and the subscript j indicates the column number. The *order* of a matrix is shown as m x n, where m is the total number of rows and n is the total number of columns. For example, consider the matrix S from Section 5.2. The order of S is 2 x 2 and the elements of S are $s_{1,1}$, $s_{1,2}$, $s_{2,1}$, and $s_{2,2}$. Placing this into a different format to show the rows and columns:

(5.11)
$$S = \begin{vmatrix} s_{1,1} & s_{1,2} \\ s_{2,1} & s_{2,2} \end{vmatrix}$$

S is then a row-column matrix; in this case a 2 row, 2 column matrix. Similarly, using the example in Section 5.2, R is a column matrix (one column with two rows):

(5.12)

$$R = \begin{vmatrix} r_1 \\ r_2 \end{vmatrix}$$

and C is a column matrix (one column with two rows):

(5.13)

$$C = \begin{vmatrix} c_1 \\ c_2 \end{vmatrix}$$

Equation 5.3 may then be re-written as:

(5.14)

$$\begin{vmatrix} r_1 \\ r_2 \end{vmatrix} = \begin{vmatrix} s_{1,1} & s_{1,2} \\ s_{2,1} & s_{2,2} \end{vmatrix} \bullet \begin{vmatrix} c_1 \\ c_2 \end{vmatrix}$$

In Equation 5.14, one would say that S *pre-multiplies*, or *left-multiplies*, C. Similarly, C *post-multiplies* or *right-multiplies* S. R is the product of S and C.

A series of operations may then be performed on matrices, summarized here as a series of theorems. Some additional definitions useful in understanding the theorems also are provided.

Theorem 5.1. Multiplication of a matrix by a constant. Let S be a matrix. If M is a new matrix equal to bS, where b is any constant, then M is a matrix of the same order as S (i.e. the same number of rows and columns), and with each element of M equal to the equivalent element of S multiplied by b. For example M = bS may be shown as:

$$\begin{vmatrix} m_{1,1} & m_{1,2} \\ m_{2,1} & m_{2,2} \end{vmatrix} = b \bullet \begin{vmatrix} s_{1,1} & s_{1,2} \\ s_{2,1} & s_{2,2} \end{vmatrix} = \begin{vmatrix} bs_{1,1} & bs_{1,2} \\ bs_{2,1} & bs_{2,2} \end{vmatrix}$$

Theorem 5.2. Identity. Two matrices are identical only if all elements with the same row and column number are the same in the two matrices. The matrix A equals the matrix B only if $a_{i,j} = b_{i,j}$ for all values of i and j. Clearly, the order of A and B must be the same.

Theorem 5.3. Symmetry. For two matrices A and B, if the relationship A = B holds, so does the relationship B = A. It is necessary (from Theorem 5.2) that the orders of A and B be the same.

Theorem 5.4. Transitivity. For three matrices A, B and C, if the relationship A = B holds, and the relationship B = C holds, so does the relationship A = C. It is necessary (from Theorem 5.2) that the orders of A, B and C be the same.

Theorem 5.5. Addition of matrices. For two matrices A and B, the sum A + B is obtained by adding the elements of A and B with identical row and column number. The result is a third matrix, C, with the same order as A and B. It is necessary that the orders of A and B be the same. In general, the element $c_{i,j}$ of the matrix C will be:

(5.15) $$c_{i,j} = a_{i,j} + b_{i,j}$$

The subtraction of matrices is governed by the same relationship. If C = A – B:

(5.16) $$c_{i,j} = a_{i,j} - b_{i,j}$$

Example 5.2. Consider two matrices A and B:

$$A = \begin{vmatrix} 2 & 3 \\ 2 & 7 \end{vmatrix} \qquad B = \begin{vmatrix} 1 & 4 \\ 3 & 6 \end{vmatrix}$$

If C = A + B:

$$C = \begin{vmatrix} 2+1 & 3+4 \\ 2+3 & 7+6 \end{vmatrix} = \begin{vmatrix} 3 & 7 \\ 5 & 13 \end{vmatrix}$$

Theorem 5.6. Commutivity. The following relationship holds for all matrices of equal order:

(5.17) $$A + B = B + A$$

Theorem 5.7. The associative rule. The following relationship holds for all matrices of equal order:

$$(A + B) + C = A + (B + C) \tag{5.18}$$

Matrix Definition 5.1. The zero matrix. This is a matrix in which all elements are zero. For example, if B is the zero matrix, $b_{i,j} = 0$ for all elements. From Theorem 5.5, if B is the zero matrix, then A + B = A.

Matrix Definition 5.2. The unit matrix. This is a *square matrix* (i.e. a matrix with the same number of rows and columns) whose elements on the *diagonal* (i.e. elements in which i=j) are 1, and whose other (*off-diagonal*) elements are 0. This is also called the *identity matrix*, I. If the unit matrix is A and has 2 rows and columns:

$$A = \begin{vmatrix} 1 & 0 \\ 0 & 1 \end{vmatrix}$$

Multiplication of any matrix, C, by the identity matrix yields the original matrix, C.

Theorem 5.8. Multiplication of matrices. Consider two matrices A and B. Left-multiplication of B by A (i.e. AB) may occur if the number of rows of B and the number of columns of A are equal. If C equals AB, then the elements of C are found from:

$$c_{i,j} = \Sigma a_{i,k} b_{k,j} \tag{5.19}$$

where the summation is over all values of k from 1 to n (the number of columns in A, which also equals the number of rows, m, in B).

Theorem 5.9. Associative rule of multiplication. The following relationship holds for all matrices capable of multiplication:

$$(AB)C = A\,(BC) \tag{5.20}$$

Theorem 5.10. Distributive rule of multiplication. The following relationships hold for all matrices capable of multiplication:

(5.21a) $$A(B+C) = AB+AC$$

(5.21b) $$(A+B)C = AC+BC$$

Theorem 5.11. Commutivity of multiplication. It is not true in general that AB = BA. There are some matrices for which this is the case, in which case it is said that these matrices *commute with one another*.

Example 5.3. Consider two matrices A and B, identical to those used in Example 5.2:

$$A = \begin{vmatrix} 2 & 3 \\ 2 & 7 \end{vmatrix} \quad B = \begin{vmatrix} 1 & 4 \\ 3 & 6 \end{vmatrix}$$

If C = AB:

$$C = \begin{vmatrix} 2\bullet1+3\bullet3 & 2\bullet4+3\bullet6 \\ 2\bullet1+7\bullet3 & 2\bullet4+7\bullet6 \end{vmatrix} = \begin{vmatrix} 11 & 26 \\ 23 & 50 \end{vmatrix}$$

Matrix Definition 5.3. The transpose. This is a matrix in which all elements i,j are switched with the elements j,i. The transpose of a matrix A is shown as A^T. The elements of A^T are:

(5.22) $$a_{i,j}{}^{T} = a_{j,i}$$

Note that the diagonal elements of the transpose are the same as the diagonal elements of the original matrix. The off-diagonal elements, however, have been switched, with the i and the j subscripts transposed. This may be seen below in Example 5.4.

Example 5.4. Consider the matrix A given by:

$$A = \begin{vmatrix} 2 & 3 \\ 1 & 7 \end{vmatrix}$$

Then:

$$A^T = \begin{vmatrix} 2 & 1 \\ 3 & 7 \end{vmatrix}$$

Matrix Definition 5.4. The inverse matrix. Consider a matrix A. If a matrix B exists such that the product of A and B yields the unit or identity matrix, I, this matrix B is the inverse of A. It is shown as A^{-1}. If $AA^{-1} = I$, A^{-1} is the *right inverse* of A. If $A^{-1}A = I$, A^{-1} is the left inverse of A. A given matrix, C, may have either a left inverse, or a right inverse, or both.

5.4. Augmented Matrices and Gauss-Jordan Elimination

The tools are now available to begin the task of unfolding environmental spectra from response data given by a response matrix, R, and from a sensitivity matrix, S. Note from Section 5.1 that R is a *column matrix* (m rows and 1 column) and S has m rows and n columns. For the example here, both m and n are 2. A new matrix may be formed by inserting the response matrix into the n+1 column of the sensitivity matrix to yield a matrix of order m x n+1. This new matrix is an *augmented matrix*, since the sensitivity matrix has been augmented by the response matrix. Using the problem in Equations 5.1 through 5.3, this new matrix will be:

(5.23)
$$S^* = \begin{vmatrix} s_{1,1} & s_{1,2} & r_1 \\ s_{2,1} & s_{2,2} & r_2 \end{vmatrix}$$

where S* is the symbol used here for the augmented matrix of S.

If R = SC as shown in Equation 5.3, the solution to the matrix C may be found directly from manipulation of the augmented matrix. Specifically, if the first two columns of the augmented matrix can be reduced to the identity matrix, the solution to C will appear in the third column (the column that produced the augmentation). This method of solution for the unknown matrix, C, is related directly to back-elimination, although in matrix form, and is known as the *Gauss-Jordan method* [5]. The general procedure for accomplishing the task relies on the fact that the final solution is not changed by dividing or multiplying all elements of any of the rows by a common constant. It also is not changed by multiplying or dividing all elements of a row by a common constant and then adding or subtracting a row from any other row. In other words, any process of multiplication, division, subtraction or addition of all elements in a row by a constant, or all elements in a row by a multiple of another row, will not change the final answer. The general procedure is as follows:

1. Divide all elements of the ith row by $s_{i,1}$. This produces a 1 as the first element in each row:

(5.24)
$$S^* = \begin{vmatrix} s_{1,1}/s_{1,1} & s_{1,2}/s_{1,1} & r_1/s_{1,1} \\ s_{2,1}/s_{2,1} & s_{2,2}/s_{2,1} & r_2/s_{2,1} \end{vmatrix}$$

2. Subtract the first row from all subsequent rows. An element in the first row is subtracted from the corresponding element (in the same column) in the other row:

(5.25)
$$= \begin{vmatrix} 1 & s_{1,2}/s_{1,1} & r_1/s_{1,1} \\ 1-1 & s_{2,2}/s_{2,1} - s_{1,2}/s_{1,1} & r_2/s_{2,1} - r_1/s_{1,1} \end{vmatrix}$$

3. Create a 1 for the element of the second row and second column by dividing all elements in this row by that element:

(5.26)
$$= \begin{vmatrix} 1 & s_{1,2}/s_{1,1} & r_1/s_{1,1} \\ 0 & (s_{2,2}/s_{2,1} - s_{1,2}/s_{1,1})/(s_{2,2}/s_{2,1} - s_{1,2}/s_{1,1}) & (r_2/s_{2,1} - r_1/s_{1,1})/(s_{2,2}/s_{2,1} - s_{1,2}/s_{1,1}) \end{vmatrix}$$

4. Create a 0 for the element of the first row and second column by multiplying the second row by this element and subtracting that row from the first row:

(5.27a)
$$= \begin{vmatrix} 1 & s_{1,2}/s_{1,1} - s_{1,2}/s_{1,1} & r_1/s_{1,1} - (s_{1,2}/s_{1,1})(r_2/s_{2,1} - r_1/s_{1,1})/(s_{2,2}/s_{2,1} - s_{1,2}/s_{1,1}) \\ 0 & 1 & (r_2/s_{2,1} - r_1/s_{1,1})/(s_{2,2}/s_{2,1} - s_{1,2}/s_{1,1}) \end{vmatrix}$$

or

(5.27b)
$$= \begin{vmatrix} 1 & 0 & r_1/s_{1,1} - (s_{1,2}/s_{1,1})(r_2/s_{2,1} - r_1/s_{1,1})/(s_{2,2}/s_{2,1} - s_{1,2}/s_{1,1}) \\ 0 & 1 & (r_2/s_{2,1} - r_1/s_{1,1})/(s_{2,2}/s_{2,1} - s_{1,2}/s_{1,1}) \end{vmatrix}$$

5. The solution matrix, C, then is in the third column of S*, or:

(5.28) $c_1 = r_1/s_{1,1} - (s_{1,2}/s_{1,1})(r_2/s_{2,1} - r_1/s_{1,1})/(s_{2,2}/s_{2,1} - s_{1,2}/s_{1,1})$

$= (r_1 - (s_{1,2}((s_{2,1}r_1/s_{1,1}) - r_2) / ((s_{2,1}s_{1,2}/s_{1,1}) - s_{2,2})))/s_{1,1}$

(5.29) $c_2 = (r_2/s_{2,1} - r_1/s_{1,1})/(s_{2,2}/s_{2,1} - s_{1,2}/s_{1,1})$

$= ((s_{2,1}r_1/s_{1,1}) - r_2) / ((s_{2,1}s_{1,2}/s_{1,1}) - s_{2,2})$

Note that these solutions for the elements of the concentration matrix, C, are the same as Equations 5.8 and 5.9, which were obtained by back-elimination. A sample calculation may be seen in Example 5.5 on the next page.

5.5. Determinants, Co-Factors, Minors and Inverses

A second, and often more computationally efficient, way to find the solution matrix in Section 5.4 is to find the inverse of the sensitivity matrix. Remember that R = SC. Left-multiplying both sides by the inverse of S (i.e. S^{-1}) yields $S^{-1}R = S^{-1}SC$. $S^{-1}S$, however, is simply the identity matrix, so $S^{-1}SC$ is simply IC or C. This leaves $C = S^{-1}R$. Our task, then, is to find the inverse of S. This inverse may be found from the following general relationship:

$$(5.30)\quad S^{-1} = (1/\Delta S) \begin{vmatrix} CF_{1,1} & CF_{1,2} & CF_{1,3} \\ CF_{2,1} & CF_{2,2} & CF_{2,3} \\ CF_{3,1} & CF_{3,2} & CF_{3,3} \end{vmatrix}^T$$

$$= (1/\Delta S) \begin{vmatrix} CF_{1,1} & CF_{2,1} & CF_{3,1} \\ CF_{1,2} & CF_{2,2} & CF_{3,2} \\ CF_{1,3} & CF_{2,3} & CF_{3,3} \end{vmatrix}$$

The quantity ΔS is the *determinant* of the matrix S [6] and the matrix C contains the *co-factors* of the matrix S as described below. Note that the *transpose* of C has been taken.

Example 5.5. Three populations are exposed to three pollutants at concentrations of c_1, c_2 and c_3. These three populations are being used as biological indicators or *biomarkers* of exposure (i.e. as measuring devices for the concentrations in the environment). The sensitivity and response matrices are:

$$S = \begin{vmatrix} 1 & 0 & 1 \\ 2 & 2 & 0 \\ 0 & 3 & 1 \end{vmatrix} \qquad R = \begin{vmatrix} 2 \\ 4 \\ 4 \end{vmatrix}$$

The augmented matrix is:

$$S^* = \begin{vmatrix} 1 & 0 & 1 & 2 \\ 2 & 2 & 0 & 4 \\ 0 & 3 & 1 & 4 \end{vmatrix}$$

Following the procedure in this section, the following transformations are made to produce the identity matrix in the first three columns. First, divide all elements of the second row by 2 and all elements of the third row by 3:

$$S^* = \begin{vmatrix} 1 & 0 & 1 & 2 \\ 1 & 1 & 0 & 2 \\ 0 & 3 & 1 & 4 \end{vmatrix} = \begin{vmatrix} 1 & 0 & 1 & 2 \\ 0 & 1 & -1 & 0 \\ 0 & 1 & 1/3 & 4/3 \end{vmatrix}$$

Subtract the second row from the third; divide the third by 4/3.

$$= \begin{vmatrix} 1 & 0 & 1 & 2 \\ 0 & 1 & -1 & 0 \\ 0 & 0 & 4/3 & 4/3 \end{vmatrix} = \begin{vmatrix} 1 & 0 & 1 & 2 \\ 0 & 1 & -1 & 0 \\ 0 & 0 & 1 & 1 \end{vmatrix}$$

Add the third row to the second; subtract the third row from the first

$$= \begin{vmatrix} 1 & 0 & 0 & 1 \\ 0 & 1 & 0 & 1 \\ 0 & 0 & 1 & 1 \end{vmatrix} \qquad \text{Therefore} \qquad C = \begin{vmatrix} 1 \\ 1 \\ 1 \end{vmatrix}$$

Matrix Definition 5.5. For the matrix S, the minor of the element $s_{i,j}$ is the matrix remaining when the ith row and j^{th} column of S are removed. For an m x n matrix, the result is an (m-1) x (n-1) matrix corresponding to each element of the matrix of minors, M. The order of M is the same as the order of S.

In this section, we will let S be a 3 x 3 matrix. There will, therefore, be 9 minors (one for each of the 9 elements of S), with each minor being a 2 x 2 matrix.

Example 5.6. Consider the matrix, S, from Example 5.5.

$$S = \begin{vmatrix} 1 & 0 & 1 \\ 2 & 2 & 0 \\ 0 & 3 & 1 \end{vmatrix}$$

There are 9 minors for this matrix, given as:

$$M_{1,1} = \begin{vmatrix} 2 & 0 \\ 3 & 1 \end{vmatrix} \quad M_{1,2} = \begin{vmatrix} 2 & 0 \\ 0 & 1 \end{vmatrix} \quad M_{1,3} = \begin{vmatrix} 2 & 2 \\ 0 & 3 \end{vmatrix}$$

$$M_{2,1} = \begin{vmatrix} 0 & 1 \\ 3 & 1 \end{vmatrix} \quad M_{2,2} = \begin{vmatrix} 1 & 1 \\ 0 & 1 \end{vmatrix} \quad M_{2,3} = \begin{vmatrix} 1 & 0 \\ 0 & 3 \end{vmatrix}$$

$$M_{3,1} = \begin{vmatrix} 0 & 1 \\ 2 & 0 \end{vmatrix} \quad M_{3,2} = \begin{vmatrix} 1 & 1 \\ 2 & 0 \end{vmatrix} \quad M_{3,3} = \begin{vmatrix} 1 & 0 \\ 2 & 2 \end{vmatrix}$$

Matrix Definition 5.6. The *co-factor matrix*. From the minors, one may construct a matrix of co-factors consisting of the elements $C_{i,j}$. Since there is one co-factor for each minor, the order of the co-factor matrix CF is the same as for the original matrix S (e.g. 3 x 3 for the matrix S in Example 5.6). The elements of CF are given by:

$$CF_{i,j} = (-1)^{i+j} \Delta M_{i,j} \qquad (5.31)$$

where $\Delta M_{i,j}$ is the determinant of the i,j element of the minor matrix, M.

Matrix Definition 5.7. The *determinant* [6]. The determinant of a matrix A is a single number, shown variously (and equivalently) as detA, ΔA or |A|. For a square 2 x 2 matrix, it is given as:

(5.32) $$\Delta A = a_{1,1}a_{2,2} - a_{1,2}a_{2,1}$$

For a square 3 x 3 matrix, it is given as:

(5.33) $$\Delta A = a_{1,1}a_{2,2}a_{3,3} - a_{1,1}a_{2,3}a_{3,2} + a_{1,2}a_{2,3}a_{3,1} - a_{1,2}a_{2,1}a_{3,3} + a_{1,3}a_{2,1}a_{3,2} - a_{1,3}a_{2,2}a_{3,1}$$

Example 5.7. Consider the matrix, M, from Example 5.6. There are 9 elements for the co-factor matrix, given as:

$CF_{1,1} = (-1)^{1+1}(2-0) = 2$

$CF_{1,2} = (-1)^{1+2}(2-0) = -2$

$CF_{1,3} = (-1)^{1+3}(6-0) = 6$

$CF_{2,1} = (-1)^{2+1}(0-3) = 3$

$CF_{2,2} = (-1)^{2+2}(1-0) = 1$

$CF_{2,3} = (-1)^{2+3}(3-0) = -3$

$CF_{3,1} = (-1)^{3+1}(0-2) = -2$

$CF_{3,2} = (-1)^{3+2}(0-2) = 2$

$CF_{3,3} = (-1)^{3+3}(2-0) = 2$

Equation 5.33 may be obtained by the process used to create the matrix of minors through the process known as *expansion by the elements of rows or columns.* First, select any single row or column of the matrix A for which the determinant is being calculated (it does not matter which you select, the determinant calculated will be the same!). For this example, we will select the first row consisting of the elements $a_{1,1}$, $a_{1,2}$ and $a_{1,3}$. The determinant of A is:

(5.34) $$\Delta A = a_{1,1}\Delta A_1 - a_{1,2}\Delta A_2 + a_{1,3}\Delta A_3$$

where ΔA_1 is the determinant of the matrix remaining when the first row and first column of A are removed, ΔA_2 is the determinant of the matrix remaining when the first row and second column of A are removed, and ΔA_3 is the determinant of the matrix remaining when the first row and third column of A are removed. This process can be expanded to any initial matrix A regardless of order, eventually reducing the calculation of the determinant to a linear combination of the determinants of 2 x 2 matrices.

For the 3 x 3 matrix A, it may be seen that:

(5.35) $\Delta A_1 = a_{2,2}a_{3,3} - a_{2,3}a_{3,2}$

(5.36) $\Delta A_2 = a_{2,1}a_{3,3} - a_{2,3}a_{3,1}$

(5.37) $\Delta A_3 = a_{2,1}a_{3,2} - a_{2,2}a_{3,1}$

Substituting Equations 5.35 through 5.37 into Equation 5.34 yields:

(5.38) $\Delta A = a_{1,1}(a_{2,2}a_{3,3} - a_{2,3}a_{3,2}) - a_{1,2}(a_{2,1}a_{3,3} - a_{2,3}a_{3,1}) + a_{1,3}(a_{2,1}a_{3,2} - a_{2,2}a_{3,1})$

which is identical to Equation 5.33.

Having found the determinant, minors and co-factors associated with the sensitivity matrix S, it then is possible to use Equation 5.30 to find the inverse of S. Given the response matrix, R, the concentration matrix may be found from:

(5.39) $$C = S^{-1}R$$

A complete example of the process may be seen through Examples 5.8 on the following pages. Note that the solution to the concentration matrix is the same as was obtained using the augmented matrix method in Example 5.5, as expected.

5.6. Applications

This final section examines several representative applications of the methods in this chapter. They by no means cover all of the applications, but give a sense of the range of problems in environmental risk assessment that might be dealt with through matrix methods.

5.6.1. System Equilibrium

Figure 5.1 shows an environmental system consisting of 3 compartments with first order rate constants between all pairs of compartments as well as rates into each compartment from outside the system and rates out of each compartment to outside the system. The general form of the differential equation for any compartment is:

(5.40) $$dN_i(t)/dt = R_i + \Sigma\lambda_{ji}N_j(t) - \Sigma\lambda_{ij}N_i(t) - \lambda_{i,out}N_i(t)$$

Example 5.8. We will use the same sensitivity matrix, S, as in Example 5.6:

$$S = \begin{vmatrix} 1 & 0 & 1 \\ 2 & 2 & 0 \\ 0 & 3 & 1 \end{vmatrix}$$

The determinant of this may be calculated from Equation 5.38 as:

$$\Delta s = 1 \bullet (2 \bullet 1 - 0 \bullet 3) - 0 \bullet (2 \bullet 1 - 0 \bullet 0) + 1 \bullet (2 \bullet 3 - 2 \bullet 0) = 8$$

The transpose of the co-factor matrix was found in Example 5.7 to be:

$$CF^T = \begin{vmatrix} 2 & 3 & -2 \\ -2 & 1 & 2 \\ 6 & -3 & 2 \end{vmatrix}$$

Therefore, using Equation 5.30 and Theorem 5.1:

$$S^{-1} = \begin{vmatrix} 2/8 & 3/8 & -2/8 \\ -2/8 & 1/8 & 2/8 \\ 6/8 & -3/8 & 2/8 \end{vmatrix}$$

Finally, C is found from left-multiplication of the response matrix R (from Example 5.5) by S^{-1}:

$$C = \begin{vmatrix} 2/8 & 3/8 & -2/8 \\ -2/8 & 1/8 & 2/8 \\ 6/8 & -3/8 & 2/8 \end{vmatrix} \begin{vmatrix} 2 \\ 4 \\ 4 \end{vmatrix} = \begin{vmatrix} 1 \\ 1 \\ 1 \end{vmatrix}$$

As expected, this is the same solution as obtained with the augmented matrix method.

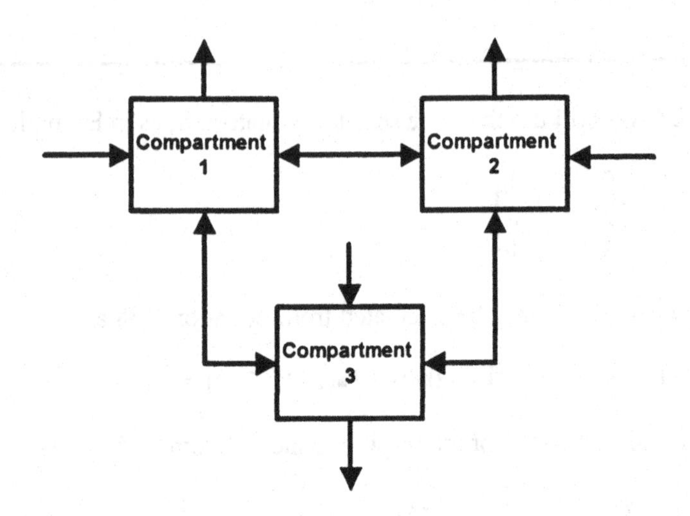

Figure 5.1. A three-compartment, open, environmental system with flows into the system from outside and out of the system from inside.

In many cases, regulators are interested in exposures that will occur once a system reaches equilibrium, since that is when exposures are highest. The argument is that, if a people still are protected against unreasonable risk when the system is at equilibrium, they will have been protected at all times prior to the system equilibrating. At equilibrium. Equation 5.40 becomes:

(5.41) $$0 = R_i + \Sigma\lambda_{ji}N_j - \Sigma\lambda_{ij}N_i - \lambda_{i,out}N_i$$

where N_i is the amount in compartment i at equilibrium, the first summation is over all values of j not equal to i, and the second summation is over all values of i not equal to i. Further, since there is no quantity λ_{ii} (the material cannot move from compartment i to compartment i), we will substitute here $\lambda_{i,out}$ for λ_{ii}. Equation 5.41 then becomes:

(5.42) $$0 = R_i + \Sigma\lambda_{ji}N_j - \Sigma\lambda_{ij}N_i$$

where the first summation is over all values of j not equal to i, and the second summation is over all values of i. The rate constants between compartments, and from a compartment to outside the system, may be written as a matrix, L:

$$L = \begin{vmatrix} -(\lambda_{11}+\lambda_{12}+\lambda_{13}) & \lambda_{12} & \lambda_{13} \\ \lambda_{21} & -(\lambda_{21}+\lambda_{22}+\lambda_{23}) & \lambda_{23} \\ \lambda_{31} & \lambda_{32} & -(\lambda_{31}+\lambda_{32}+\lambda_{33}) \end{vmatrix}$$

The equilibrium amounts in each of the three compartments may also be written as a matrix:

$$N = \begin{vmatrix} N_1 \\ N_2 \\ N_3 \end{vmatrix}$$

The system of equations given by Equation 5.42 may then be written as:

$$\begin{vmatrix} 0 \\ 0 \\ 0 \end{vmatrix} = \begin{vmatrix} R_1 \\ R_2 \\ R_3 \end{vmatrix} + \begin{vmatrix} -\Sigma\lambda_{1j} & \lambda_{12} & \lambda_{13} \\ \lambda_{21} & -\Sigma\lambda_{2j} & \lambda_{23} \\ \lambda_{31} & \lambda_{32} & -\Sigma\lambda_{3j} \end{vmatrix}^T \times \begin{vmatrix} N_1 \\ N_2 \\ N_3 \end{vmatrix}$$

or

(5.43) $$[0] = [R] + [L]^T[N]$$

where 0 is the *null matrix* (i.e. a matrix with all elements equal 0). Rearranging Equation 5.43:

(5.44) $$([0] - [R]) = [L]^T[N]$$

(5.45) $$[L^T]^{-1}([0] - [R]) = [L^T]^{-1}[L]^T[N]$$

where $[L^T]^{-1}$ is the inverse of the transpose of [L]. Since $[L^T]^{-1}[L^T]$ is the identity matrix, and since $([0] - [R]) = -[R]$:

(5.46) $$N = -[L^T]^{-1}[R]$$

5.6.2. Fields of Aggregate Exposure and Risk

Chapter 1 introduced the concept of fields and their relationship to geographic regions. Imagine three such two-dimensional fields superimposed over the same geographic region. The region is divided into a grid, with different concentrations of a pollutant in the air, water and food in each grid block. The grid blocks themselves are indexed as i,j with i being the row number (from north to south, looking down onto the geographic region from the bird's eye view) and j being the column number (from west to east). The grid system is shown in Figure 5.2. The concentration of the pollutant in air, water and food in each grid block then is $C^a_{i,j}$, $C^w_{i,j}$ and $C^f_{i,j}$, respectively.

Aggregate exposure is defined as the total rate of intake of a pollutant into the body by all exposure routes, divided by the body mass [7]. Aggregate exposure then has units of mg/kg-day if the concentration is in mg/L, the rate of inhalation or ingestion has units of L/day, and the body weight has units of kg. In this example, there are three such routes (inhalation of air, ingestion of water and ingestion of soil). For a given grid block, the aggregate exposure, $AE_{i,j}$, then is:

$$(5.47) \qquad AE_{i,j} = (IR_a C^a_{i,j} + IR_w C^w_{i,j} + IR_f C^f_{i,j})$$

where IR_j is the intake rate of the jth environmental medium. The risk is assumed proportional to the aggregate exposure, with the proportionality constant being shown as a *slope factor*, SF, having units of $(mg/kg\text{-}day)^{-1}$. The risk in a grid block then may be calculated from:

1,1	1,2	1,3
2,1	2,2	2,3
3,1	3,2	3,3

Figure 5.2. A geographic region divided into grid blocks representing a matrix in which exposures to a pollutant in air, water and food may occur.

$$(5.48) \qquad R_{i,j} = SF \bullet AE_{i,j} = SF \bullet (IR_a C^a_{i,j} + IR_w C^w_{i,j} + IR_f C^f_{i,j})$$

Equation 5.48 may then be formulated as a series of matrix operations:

$$(5.49) \qquad [R] = SF \bullet [AE] = SF \bullet (IR_a \bullet [C^a] + IR_w \bullet [C^w] + IR_f \bullet [C^f])$$

5.6.3. Watersheds and Run-off

In this final application, we consider a watershed in which rainfall onto any one grid-block in a geographic region such as the one in Figure 5.2 may flow to any other grid-block in the region. We first must reformat the matrix associated with Figure 5.2. In that figure, there are 9 gridblocks. We create a new matrix of order 9 x 9 (for a total of 81 elements) given by [F], where F stands for "fraction". This is done by assigning each of the gridblocks in Figure 5.2 to a new index k in the following manner:

- Gridblock 1,1 in Figure 5.2 is assigned k=1
- Gridblock 1,2 in Figure 5.2 is assigned k=2
- Gridblock 1,3 in Figure 5.2 is assigned k=3
- Gridblock 2,1 in Figure 5.2 is assigned k=4
- Gridblock 2,2 in Figure 5.2 is assigned k=5
- Gridblock 2,3 in Figure 5.2 is assigned k=6
- Gridblock 3,1 in Figure 5.2 is assigned k=7
- Gridblock 3,2 in Figure 5.2 is assigned k=8
- Gridblock 3,3 in Figure 5.2 is assigned k=9

The elements of this new matrix, $f_{i,j}$, represent the fraction of water falling into gridblock i which then moves to grid-bock j. For example, $f_{3,6}$ is the fraction of water falling into the gridblock assigned k=3 above (i.e. gridblock 1,3 in Figure 5.2) which then moves into the gridblock assigned k=6 above (i.e. gridblock 2,3 in Figure 5.2). The diagonal elements ($f_{i,i}$) represent the fraction of the rain falling into gridblock i that remains in gridblock I (which is simply 1 minus the sum of the fractions going from gridblock i to all other gridblocks j).

Suppose the rate of rainfall onto gridblock j is RF_j. The total rate of water entering gridblock j, R_j, taking into account both the water remaining in j after falling in j and the water moving to j after falling in i is:

(5.50) $$R_j = RF_j f_{j,j} + \Sigma\, RF_i f_{i,j}$$

where the summation is over all values of i except i = j. If the summation is over all values of i including i = j:

(5.51) $$R_j = \Sigma\, RF_i f_{i,j}$$

To formulate this properly as a matrix problem, the order of the terms in the summation must be reversed to allow multiplication of [F] and [RF]:

(5.52) $$R_j = \Sigma f_{i,j} RF_j$$

We then define the rainfall matrix [RF] as a 9 x 1 column matrix (9 rows and 1 column), and the fraction matrix [F] as a 9 x 9 matrix. The matrix describing the rate of water flow into each grid block then is [R], which also is a column matrix of order 9 x 1. The resulting matrix problem may then be written as:

(5.53) $$[R] = [F]^T[RF]$$

Finding the rate of flow of water into each gridblock then requires taking the transpose of [F] and multiplying this by [RF].

References

1. B. Blanchard and W. Fabrycky, *Systems Engineering and Analysis*, Prentice Hall, Upper Saddle River, New Jersey, 1998.
2. G. Knoll, *Radiation Detection and Measurement*, John Wiley and Sons, New York, 1999.
3. D. Hill and B. Kolman, *Modern Matrix Algebra*, Prentice Hall, Upper Saddle River, New Jersey, 2001.
4. H. Anton, *Elementary Linear Algebra*, John Wiley and Sons, New York, 2000.
5. R. Allenby, *Linear Algebra*, Butterworth-Heinemann, Woburn, Massachusetts, 1995.
6. L. Spence, A. Insel, L. Friedberg, *Elementary Linear Algebra: A Matrix Approach*, Prentice Hall, Upper Saddle River, New Jersey, 1999.
7. D. Crawford-Brown, *Risk-Based Environmental Decisions: Methods and Culture*, Kluwer Academic Publishers, Boston, 2000.

CHAPTER 6
Numerical Methods and Exposure-Response

6.1. Exposure-Response Relationships

Catenary systems of compartments, in which flow of a pollutant is in one direction through a chain of compartments and where the rates of movement are controlled by first-order kinetics, can be described by differential equations with analytic solutions regardless of the number of compartments (see Chapter 3). Even if flow is in several directions, analytic solutions may be found for some relatively simple cases (see Chapter 4). As the complexity of the system increases, however, with large numbers of mutually connected compartments, the ability to produce analytic solutions to the underlying differential equations increases greatly. Even if analytic solutions can be found in principle, they may involve such long derivations as to make the task impractical.

Faced with such complexity, it is necessary to seek other ways to generate predictions of the behavior of a system. The advantage of an analytic solution is that the final equation has a specific mathematical form (such as the exponential wash-out equation from Chapter 3) whose terms can be interpreted physically. In other words, the equation representing an analytic solution contains information and insights about the nature of the system being modeled. If one gives up the requirement for such conceptual insights, however, and turns only to the issue of predicting system behavior rather than fully explaining it, there is a variety of methods that may be used even if analytic solutions cannot be obtained. These methods produce estimates of the numerical values of system properties (e.g. concentrations) across spatial and/or temporal domains, and hence are called *numerical methods*.

One such set of methods involves *numerical solution* of the equations describing a system. Numerical methods have been studied extensively in both science and applied mathematics because physical systems can grow very quickly in complexity beyond the limits of available analytical methods. This chapter examines some of these methods as they are applied to solving both integrals and differential equations. As in previous chapters, the reader should view the methods discussed here as introductory, and should explore more complex (and often more reliable and accurate) methods developed in the various references cited. Still, the methods developed here will provide reasonably reliable predictions under the conditions presented in the Chapter.

We begin with an example of a compartmental system with three compartments and flow between each pair of compartments. In addition, there is a rate into the first compartment from a source. The hypothetical system, consisting of soil, water and plants is shown in Figure 6.1.

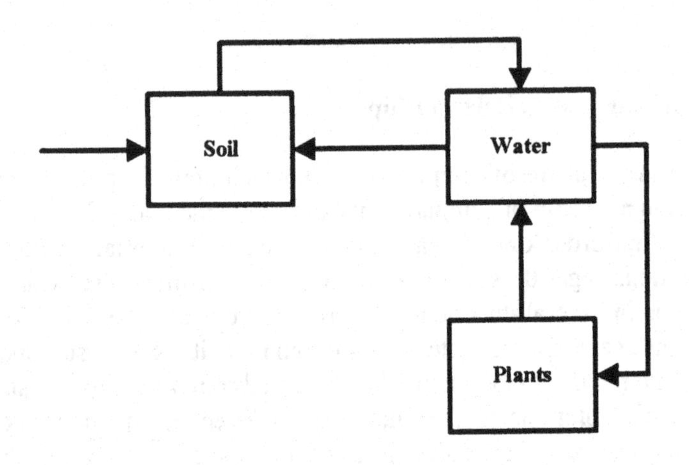

Figure 6.1. A three-compartment, open system with bi-directional flow. Note the flow into the system through the soil.

Let the rate into the soil from outside the system be given as R_{in}, and all other transfers between conmpartments be given by first-order processes. The transfer rate constants then are λ_{sw} for flow from the soil to water; λ_{ws} for flow from the water to soil; λ_{wp} for flow from the water to plants; and λ_{pw} for flow from the plants to water. The differential equations describing the system are:

$$dN_s(t)/dt = R_{in} + \lambda_{ws}N_w(t) - \lambda_{sw}N_s(t) \tag{6.1}$$

$$dN_w(t)/dt = \lambda_{sw}N_s(t) + \lambda_{pw}N_p(t) - \lambda_{ws}N_w(t) - \lambda_{wp}N_w(t) \tag{6.2}$$

$$dN_p(t)/dt = \lambda_{wp}N_w(t) - \lambda_{pw}N_p(t) \tag{6.3}$$

This system of differential equations cannot be solved via Bernoulli's methods, and solutions with Laplace transforms would be time-consuming and tedious.

Imagine now that the complexity of the problem is increased by considering movement of the pollutant into the body through ingestion of the soil, water and plants. The rate of ingestion of each environmental medium is IR_s, IR_w

and IR_p, respectively (in units, for example, of m^3 per day). The rate of movement of a pollutant into the body at any moment then is given by:

$$R_{body}(t) = C_s(t)IR_s(t) + C_w(t)IR_w(t) + C_p(t)IR_p(t) \tag{6.4}$$

where $C_s(t)$, $C_w(t)$ and $C_p(t)$ are the solutions to Equations 6.1 through 6.3, respectively at time t, each divided by the volume, V, of their respective compartments. Note that the intake rates also are a function of time, since these generally change with age [1]. If N(t) is in grams, V is in m^3, and IR is in m^3/d, R is in g/d.

Once the pollutant enters the body, it moves into the gastrointestinal tract, is absorbed into the bloodstream, and circulates between various organs before being excreted. This movement is shown in Figure 6.2.

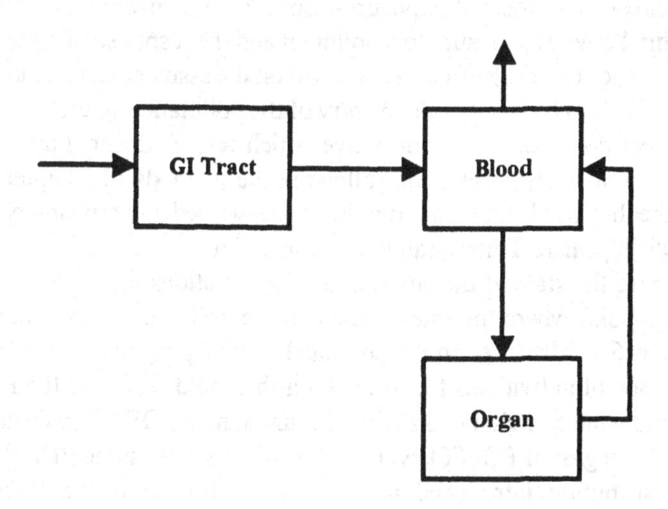

Figure 6.2. A three-compartment, open system with bi-directional flow in the body representing a pharmacokinetic model. Note the flow into the system through the environment (left-most arrow) described by Equation 6.4.

Let the rate into the body from the environment be given as $R_{body}(t)$, as described in Equation 6.4 and all other transfers be given by first-order processes. The transfer rate constants then are λ_{GIb} for flow from the GI tract to blood; λ_{bo} for flow from the blood to organ; λ_{ob} for flow from the organ to blood; and $\lambda_{b,out}$ for flow from the blood to outside the body via excretion. The differential equations describing the system then are:

$$dN_{GI}(t)/dt = R_{body}(t) - \lambda_{GIb}N_{GI}(t) \tag{6.5}$$

(6.6) $$dN_b(t)/dt = \lambda_{GIb}N_{GI}(t) + \lambda_{ob}N_o(t) - \lambda_{b,out}N_b(t) - \lambda_{bo}N_b(t)$$

(6.7) $$dN_o(t)/dt = \lambda_{bo}N_b(t) - \lambda_{ob}N_o(t)$$

This system of differential equations cannot be solved via Bernoulli's methods, and solutions with Laplace transforms again would be time-consuming and tedious since $R_{body}(t)$ is not constant in time.

To complicate the situation further, imagine that the task at hand is not simply to calculate the amount of pollutant in each compartment of the environment or of the body, but rather to estimate the probability of a health effect (e.g. cancer) in a population exposed to this environment. To perform this latter calculation, one might select from amongst two classes of models developed for such purposes: *distributed threshold* models and *state-vector* models [1]. Distributed threshold models of exposure-response relationships (i.e. models of the relationship between exposure to a pollutant and the response of a population through appearance of health effects) are based on the assumption that individuals in a population have a threshold rate of entry of the pollutant into the body below which the effect does not occur but above which it does occur. Further, each individual has a different threshold. Following the methods of Chapter 2, this variation in the threshold amongst individuals is described by a probability density function, PDF(T), where T here stands for "threshold".

Suppose the state of the environment in Equations 6.1 through 6.3 has evolved to the point where the rate of entry of the pollutant into the body is R using Equation 6.4. The fraction of individuals developing the effect will then equal the fraction of individuals for whom their threshold, T, is less than R. This fraction is simply the cumulative distribution function for PDF(T), evaluated up to R. Or, it is the integral of PDF(T) evaluated from 0 to R. Assume PDF(T) is the lognormal distribution introduced in Chapter 2. If F(R) is the fraction of individuals developing the effect in a population of individuals who each have an intake rate R of the pollutant into the body, then using the PDF shown in Equation 2.9:

(6.8) $$F(R) = \int \exp[(-(\ln T - \ln T_m)^2/(2 \bullet \ln^2 \sigma_g)] / [T \bullet \sigma_g \bullet (2\pi)^{0.5}]dT$$

where the integration is from 0 to R, T_m is the median threshold value for the population and σ_g is the geometric standard deviation of the thresholds in this population. This integral cannot be solved in closed form, and so it must be integrated numerically. The exposure-response relationship, therefore, requires use of numerical methods of integration.

More *biologically-based models* (i.e. models that incorporate more of the information about the processes occurring in the body) use the concept of a *state vector*. The body is assumed to pass through a series of states of health on the way from complete health to appearance of a disease. These transitions between states may occur either due to background events (e.g. random mutations) or due to exposure to a pollutant. The rates of transitions between states then are functions of the amount of the pollutant present in a *target organ* at any moment (i.e. the organ in which the damage must first occur to produce the effect). An example of a simple state-vector model is shown in Figure 6.3. In this model, there are three states of health (normal cells, initiated cells and cancer cells); the effect appears only in cells moved to the last state (cancer). The pollutant causes transitions upward through the chain of states by producing damage in an organ (see Figure 6.2, which shows the organ). There also are transitions backwards through the states due to repair of the damage.

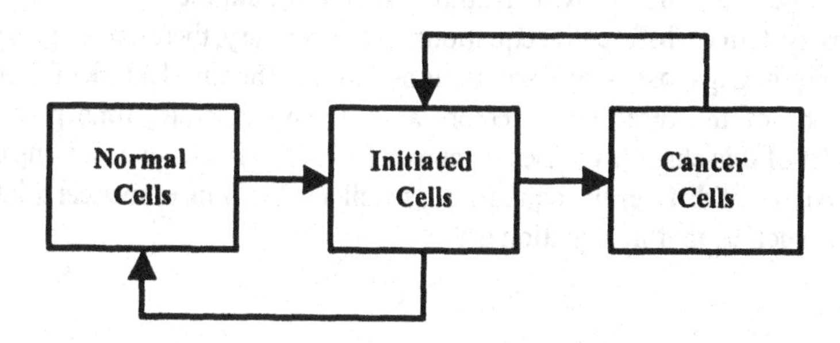

Figure 6.3. An example of a state vector model of exposure-response. Transitions occur in both directions within the chain of states of health. Some of these transitions are affected by the presence of the pollutant in the target organ.

Let the amount of the pollutant in the target organ be given as $N_o(t)$, as described by Equation 6.7 and let all other transitions between states of health be given by first-order processes. The transition rate constants then are λ_{ni} for transitions from normal to initiated cells; λ_{in} for repair backwards from initiated to normal cells; λ_{ic} for transitions from initiated to cancer cells; and λ_{ci} for repair backwards from cancer to initiated cells. The forward transition rate constants have units of probability per second per gram of pollutant in the target organ, while the backward transition rate constants have units of probability per second (the repair processes are not caused in this model by the pollutant in the target organ, and so the rate constant backwards is not a function of the amount of the pollutant in the target organ). Note that only the rates of forward transitions then are a function of the amount of the pollutant in the target organ. The differential equations describing this system are:

(6.9) $$dN_n(t)/dt = \lambda_{in}N_i(t) - N_o(t)\lambda_{ni}N_n(t)$$

(6.10) $$dN_i(t)/dt = N_o(t)\lambda_{ni}N_n(t) + \lambda_{ci}N_c(t) - \lambda_{in}N_i(t) - N_o(t)\lambda_{ic}N_i(t)$$

(6.11) $$dN_c(t)/dt = N_o(t)\lambda_{ic}N_i(t) - \lambda_{ci}N_c(t)$$

Solving this system of differential equations requires first that Equations 6.5 through 6.7 be solved (yielding the state vector for the pollutant in the body), which in turn requires that Equations 6.1 through 6.3 be solved (yielding the state vector for amounts, or concentrations, of the pollutant in the compartments of the environment). Note that this problem introduces a concepts mentioned in Chapter 1: the state of the environment affects the sate of the pollutant in the body, which in turn affects the state of health.

There are no analytic methods that are suitable for solving such a complex system of differential equations. It is necessary, therefore, to give yup the goal of producing closed-form solutions and turn to the simpler task of developing predictions of the three state vectors at different, discrete, time points. The remainder of this chapter focuses on numerical methods useful in solving this and other systems of differential equations, as well as obtaining numerical solutions to integrals such as that in Equation 6.8.

6.2. Numerical Integration

Consider the lognormal PDF that must be integrated in Equation 6.8, which is shown graphically in Figure 6.4. While the integral is difficult to obtain analytically, it is a simple matter to calculate the value of the PDF at any value of T by inserting T into the PDF. Now note that the integral of a function between two endpoints (e.g. from T equal 2 to T equal 4 in Figure 6.4) is the area under that curve between these endpoints. Any procedure that allows calculation of this area also allows integration of the PDF.

As a first approximation, consider calculating the values of the PDF at discrete points along the x-axis. For example, one might be interested in the solution to Equation 6.8 for a value of R equal to 4 (i.e. the fraction of individuals showing the effect at R equal 4, which is the fraction of people whose threshold is at or below 4). This requires integration of Equation 6.8, or Figure 6.4, between T equal 0 and T equal 4. The values of the PDF at T equal 1 and T equal 3 are 0.085 and 0.175, respectively. Assuming the value at T equal 1 applies to the entire region from T equal 0 to T equal 2, the area under the PDF in this region will be 0.085 x 2 = 0.17. Similarly, assuming the value at T equal 3 applies to the entire region from T equal 2 to T equal 4, the area under the PDF in this region will be

0.175 x 2 = 0.35. The total area under the PDF from T equal 0 to T equal 4 then is 0.35 + 0.17 = 0.52. 52% of the population will show the effect if R equals 4 mg/kg-day. Note that the result should be exactly 50% since the median is 4. The result is shown in Figure 6.5.

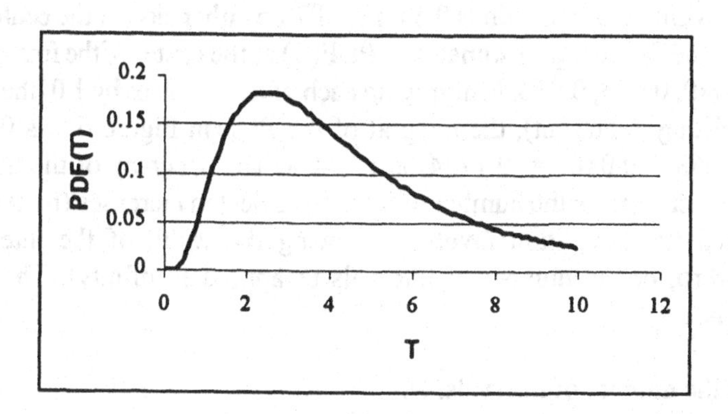

Figure 6.4. The PDF for thresholds in a population exposed to a pollutant. In this figure, the units of the threshold, T, are mg/kg-day, or the rate of entry into the body divided by the body mass (this is the *average daily rate of intake*, or ADRI, in regulatory risk assessment [2]). The median value of the threshold is 4 and the geometric standard deviation is 2.

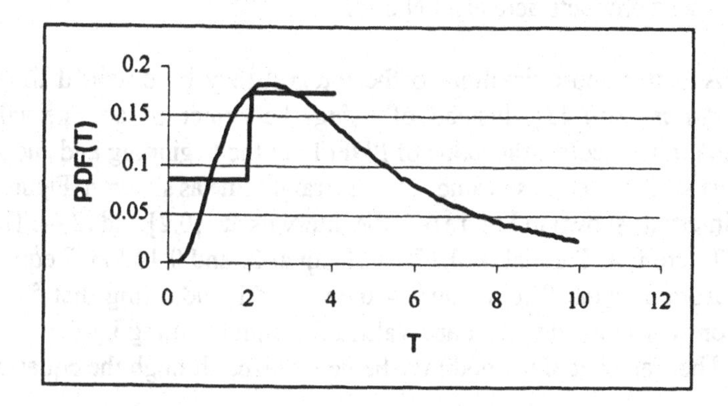

Figure 6.5. Integration of the PDF for thresholds in a population exposed to a pollutant. In this figure, the units of the threshold are mg/kg-day. The PDF is the same as in Figure 6.4. Two intervals have been selected to integrate the PDF from 0 to 4. The integral is the sum of the areas under the two rectangles.

Note that the *equivalent rectangles* in Figure 6.5 are not very good approximations of the function except at the points 1 and 3 (where they are exact). To obtain a better estimate of the integral, the number of points at which the PDF is evaluated may be increased. Increasing to 4 intervals centered at 0.5, 1.5, 2.5, 3.5, and letting the values calculated at these 4 locations approximate the values of the PDF within plus and minus 0.5 units of T on either side of the center point, one obtains the following 4 estimates of PDF(T) at the center of the four intervals: 0.0044, 0.049, 0.063, 0.056. Multiplying each of these values by 1.0 (the width of the interval they represent), the integral of the PDF in Figure 6.5 is 0.0044 + 0.049 + 0.063 + 0.056 = 0.1724 or 17.24%. The accuracy of the numerical integration increases as the number of intervals selected increases (remembering that an analytic integration involves allowing the width of the intervals to approach zero, or the number of intervals to approach infinity). The general procedure is:

1. Select the number of intervals, N.
2. Locate the center value of T for each interval.
3. Calculate PDF(T) for each of these N center values.
4. Multiply each of these values of PDF(T) by ΔT, the width of the intervals.
5. Sum these products across all N intervals.

A better approximation to the integral may be obtained through the *trapezoidal method* [3]. Instead of using the center of an interval as the approximation, calculate the value of PDF(T) at the beginning and endpoints of each interval. Connect these values with a straight line as shown in Figure 6.6. For the two intervals shown in Figure 6.6, the intervals are [0,2] and [2,4]. The values of PDF(T) are 0 at T equal 0; 0.175 at T equal 2; and 0.142 at T equal 4. The integral from T equal 0 to T equal 4 then is 0.51, indicating that 51% of the population will show the effect at a value of R equal to 4 mg/kg-day.

The trapezoidal method may be generalized through the equation:

$$(6.12) \quad \int F(x)dx = [F(x_1) + 2F(x_2) + 2F(x_3) + \ldots + F(x_N)]\,\Delta x / 2$$

where the function F(x) is integrated from point x_1 to x_N, where there are N points (values of x) at which the function F(x) is evaluated, and where Δx is the width of the intervals.

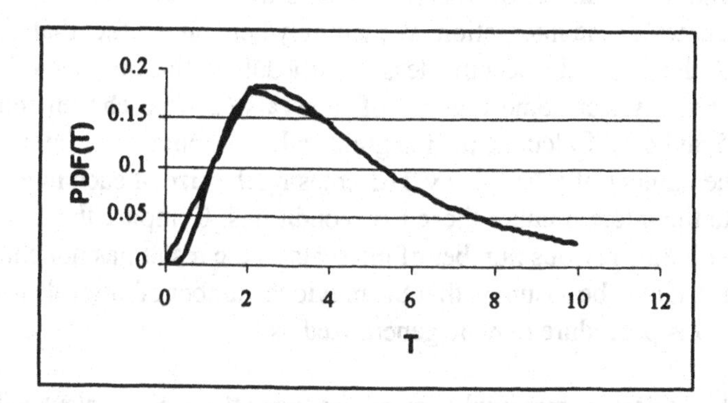

Figure 6.6. Integration of the PDF for thresholds in a population exposed to a pollutant, using the trapezoidal method. In this figure, the units of the threshold are mg/kg-day. The PDF is the same as in Figure 6.5. Two intervals have been selected to integrate the PDF from 0 to 4. The integral is the sum of the areas under the two trapezoids.

Example 6.1. A population exposed to a pollutant has an elevated probability of liver cancer. The rate at which new cases of liver cancer appear (cases per year) is given by the relationship:

$$R(t) = e^{-0.2t}t^2$$

where t is the time in years. How many new cases, N, appear during the interval from t equal 4 to t equal 8 years? To solve this problem, note that the number of new cases appearing in an interval of time is the integral of the rate of appearance, R(t), over this time interval. In other words:

$$N = \int R(t)dt = \int e^{-0.2t}t^2\,dt$$

where the integrals are from 4 to 8. Using Equation 6.12 with 5 points at 4, 5, 6, 7 and 8, and with Δt equal 1 year:

$$\int e^{-0.2t}t^2\,dt = [e^{-0.2\bullet 4}4^2 + 2\bullet e^{-0.2\bullet 5}5^2 + 2\bullet e^{-0.2\bullet 6}6^2 + 2\bullet e^{-0.2\bullet 7}7^2 + e^{-0.2\bullet 8}8^2]\bullet 1/2 = 42.2$$

For both the method of equivalent rectangles and the trapezoidal method, the accuracy of the integration increases as the number of intervals increases. How might the requisite number of intervals be determined? This will depend on the desired accuracy of the integration. We will rely here on the fact that changes in the size of the intervals become less significant as the number of intervals increases. First, select some number of intervals (2 were chosen initially for Figures 6.5 and 6.6). Calculate the integrals under this number of intervals. Then multiply the number of intervals by 2 (decreasing the size of each interval by 2). Recalculate the integral under these new conditions. Compare the result to that obtained with the previous number of intervals. If the result has not changed too dramatically, it may be assumed that the previous number of intervals was already sufficient. This procedure may be generalized as:

1. Establish a desired maximal allowed change in the integral calculated between any two selections of the number of intervals (e.g. 0.01 or 1%).

2. Select an initial number, N, for the number of intervals.

3. Calculate the integral with N intervals. This integral value is I_N.

4. Multiply N by 2.

5. Calculate the integral using 2N intervals. This new integral value is I_{2N}.

6. Calculate $(I_N-I_{2N})/I_N$.

7. If the absolute value of this ratio is greater than the maximal allowed change from Step 1 (e.g. greater than 0.01), return to Step 4 and increase N; otherwise, accept this new value of the integral.

There is no generally accepted rule as to how small the maximal allowed change should be, although values of less than 0.001 often are applied since computation time is short with modern computers.

Note that the methods discussed in this section use a constant width of the intervals. For example, the width of the interval from 0 to 2 is the same as the width of the interval from 2 to 4. While there is nothing wrong conceptually with such an approach, it can lead to inefficient computations. This can be understood by considering the trapezoidal method and Figure 6.6. The method is based on a calculation of the slope of the PDF (or whatever function is being integrated) between the endpoints of an interval. This slope is the straight line connecting those endpoints. If the slope is constant throughout the interval, the trapezoidal method will give an exact solution to the integral in that interval. If the slope changes, however, meaning the function has a non-zero second derivative throughout the interval, the trapezoidal method (and the equivalent rectangle method) will produce inaccuracies in the estimate of the integral. These inaccuracies will decrease as the size of the intervals decreases.

The size of the interval needed to produce a given level of inaccuracy (e.g. 1%) depends on the second derivative of the function being integrated. Where the second derivative is large, the slope is changing rapidly in the vicinity of that point and both methods described here will produce relatively large inaccuracies if the interval size also is large. The goal is to select intervals sufficiently small that the second derivative also is small in each of them. If a single interval size is selected for use throughout the region of integration, it will be necessary to select a size that will prove adequate even in the regions of the curve where the second derivative is large. Such small intervals, however, are not needed in regions of the curve where the second derivative is small. In these latter regions, the small intervals will produce greater accuracy than required. This is not a problem conceptually, since greater accuracy always is good, but it does increase the number of calculations that must be performed in these regions of the curve (and, therefore, the length of time required for the integration) with little benefit to the overall accuracy of the integration.

Methods have been developed to deal with this issue by examining the second derivative of the function being integrated as the integration proceeds. In regions of the integration where this second derivative is large, the size of the intervals is selected to be small. In regions of the integration where this second derivative is small, the size of the intervals is selected to be large. The result is similar levels of accuracy in each of the regions, resulting in computational efficiency. This increase in efficiency can be important in problems requiring millions of integrations. The methods allowing this selection of different interval sizes are beyond the level of this book, and so the reader is referred to other references [4].

6.3. Numerical Solutions to Differential Equations: Euler's Method

Methods similar to those introduced in Section 6.2 may be used to solve the system of differential equations introduced in Section 6.1. To understand this, it is necessary to note only that the solution to a differential equation at any value of time, t, is the integral of that equation from 0 to t added to any initial value of the quantity being estimated.

Consider a case in which Figures 6.1, 6.2 and 6.3 describe the environmental state, pharmacokinetics and exposure-response characteristics, respectively, of a situation in which a population is exposed to a pollutant. The equations are repeated here in slightly revised form to show their interconnections:

(6.13) $$dN_s(t)/dt = R_{in} + \lambda_{ws}N_w(t) - \lambda_{sw}N_s(t)$$

(6.14) $$dN_w(t)/dt = \lambda_{sw}N_s(t) + \lambda_{pw}N_p(t) - \lambda_{ws}N_w(t) - \lambda_{wp}N_w(t)$$

(6.15) $$dN_p(t)/dt = \lambda_{wp}N_w(t) - \lambda_{pw}N_p(t)$$

(6.16) $$dN_{GI}(t)/dt = N_s(t)IR_s(t)/V_s + N_w(t)IR_w(t)/V_w + N_p(t)IR_p(t)/V_p - \lambda_{GIb}N_{GI}(t)$$

(6.17) $$dN_b(t)/dt = \lambda_{GIb}N_{GI}(t) + \lambda_{ob}N_o(t) - \lambda_{b,out}N_b(t) - \lambda_{bo}N_b(t)$$

(6.18) $$dN_o(t)/dt = \lambda_{bo}N_b(t) - \lambda_{ob}N_o(t)$$

(6.19) $$dN_n(t)/dt = \lambda_{in}N_i(t) - N_o(t)\lambda_{ni}N_n(t)$$

(6.20) $$dN_i(t)/dt = N_o(t)\lambda_{ni}N_n(t) + \lambda_{ci}N_c(t) - \lambda_{in}N_i(t) - N_o(t)\lambda_{ic}N_i(t)$$

(6.21) $$dN_c(t)/dt = N_o(t)\lambda_{ic}N_i(t) - \lambda_{ci}N_c(t)$$

where V_s, V_w and V_p are the volumes of the soil, water and plant compartments, respectively. The task at hand is to calculate the number of cancer cells, which is assumed to be proportional to the risk of cancer in the population. In other words, the fraction of people developing cancer as of time t equals a constant times $N_c(t)$.

In numerical integration (Section 6.2), the value of the function being integrated was calculated at discrete points and used in the integration. For solutions to differential equations, the values of the functions sought cannot be calculated (otherwise there would be no need to solve the differential equation). What is known in this case is the slope, or derivative, of the function at discrete points, since this slope is described by the differential equations. These estimates of the slope, as well as the initial values for the system of compartments, may be used to construct the solution piecewise. The pieces are steps forward in time, and so the methods explored here use the concept of *discrete time-steps*.

In *Euler's method* [5], which is the simplest but least accurate method of numerical solutions to differential equations, one begins with the initial values for the compartments (for the environmental and pharmacokinetic models) and states of health (for the dose-response model) in the system. There are 9 such compartments and states in Equations 6.13 to 6.21. The initial values for these are $N_s(0)$, $N_w(0)$, $N_p(0)$, $N_{GI}(0)$, $N_b(0)$, $N_o(0)$, $N_n(0)$, $N_i(0)$ and $N_c(0)$; these values must be known for the method to begin.

Equations 6.13 to 6.21 may then be used to calculate the slopes, or derivatives, of each of the 9 functions at t equal 0. These are found by replacing the functions on the right–hand sides of these relationships by their respective initial values. The designation dN(t)/dt then is replaced by ΔN(t)/Δt since the method uses discrete, finite, time-steps rather than infinitesimal time steps:

$$\Delta N_s(0)/\Delta t = R_{in} + \lambda_{ws}N_w(0) - \lambda_{sw}N_s(0) \quad (6.22)$$

$$\Delta N_w(0)/\Delta t = \lambda_{sw}N_s(0) + \lambda_{pw}N_p(0) - \lambda_{ws}N_w(0) - \lambda_{wp}N_w(0) \quad (6.23)$$

$$\Delta N_p(0)/\Delta t = \lambda_{wp}N_w(0) - \lambda_{pw}N_p(0) \quad (6.24)$$

$$\Delta N_{GI}(0)/\Delta t = N_s(0)IR_s(0)/V_s + N_w(0)IR_w(0)/V_w + N_p(0)IR_p(0)/V_p - \lambda_{GIb}N_{GI}(0) \quad (6.25)$$

$$\Delta N_b(0)/\Delta t = \lambda_{GIb}N_{GI}(0) + \lambda_{ob}N_o(0) - \lambda_{b,out}N_b(0) - \lambda_{bo}N_b(0) \quad (6.26)$$

$$\Delta N_o(0)/\Delta t = \lambda_{bo}N_b(0) - \lambda_{ob}N_o(0) \quad (6.27)$$

$$\Delta N_n(0)/\Delta t = \lambda_{in}N_i(0) - N_o(0)\lambda_{ni}N_n(0) \quad (6.28)$$

$$\Delta N_i(0)/\Delta t = N_o(0)\lambda_{ni}N_n(0) + \lambda_{ci}N_c(0) - \lambda_{in}N_i(0) - N_o(0)\lambda_{ic}N_i(0) \quad (6.29)$$

$$\Delta N_c(0)/\Delta t = N_o(0)\lambda_{ic}N_i(0) - \lambda_{ci}N_c(0) \quad (6.30)$$

At t equal 0, both the value of each function, N(0), and the slope, ΔN(0)/Δt, are known (the slope is simply the value of the right-hand side of Equations 6.22 through 6.30). Assume this same slope will continue to hold true over some finite time-step, Δt. This is only approximately valid since, as is evident from Equations 6.13 through 6.21, the slopes of the functions in fact change with time. Still, so long as Δt is not very large, the approximation will be valid. Beginning with N(0), the amount in the compartment or state at the end of the first time-step Δt will be:

$$N(0+\Delta t) = N(\Delta t) = N(0) + [\Delta N(0)/\Delta t]\Delta t \quad (6.31)$$

For Equations 6.22 through 6.30 this yields:

$$N_s(\Delta t) = N_s(0) + [R_{in} + \lambda_{ws}N_w(0) - \lambda_{sw}N_s(0)]\Delta t \quad (6.32)$$

$$N_w(\Delta t) = N_w(0) + [\lambda_{sw}N_s(0) + \lambda_{pw}N_p(0) - \lambda_{ws}N_w(0) - \lambda_{wp}N_w(0)]\Delta t \quad (6.33)$$

(6.34) $$N_p(\Delta t) = N_p(0) + [\lambda_{wp}N_w(0) - \lambda_{pw}N_p(0)]\Delta t$$

(6.35) $$N_{GI}(\Delta t) = N_{GI}(0) + [N_s(0)IR_s(0)/V_s + N_w(0)IR_w(0)/V_w + N_p(0)IR_p(0)/V_p - \lambda_{GIb}N_{GI}(0)]\Delta t$$

(6.36) $$N_b(\Delta t) = N_b(0) + [\lambda_{GIb}N_{GI}(0) + \lambda_{ob}N_o(0) - \lambda_{b,out}N_b(0) - \lambda_{bo}N_b(0)]\Delta t$$

(6.37) $$N_o(\Delta t) = N_o(0) + [\lambda_{bo}N_b(0) - \lambda_{ob}N_o(0)]\Delta t$$

(6.38) $$N_n(\Delta t) = N_n(0) + [\lambda_{in}N_i(0) - N_o(0)\lambda_{ni}N_n(0)]\Delta t$$

(6.39) $$N_i(\Delta t) = N_i(0) + [N_o(0)\lambda_{ni}N_n(0) + \lambda_{ci}N_c(0) - \lambda_{in}N_i(0) - N_o(0)\lambda_{ic}N_i(0)]\Delta t$$

(6.40) $$N_c(\Delta t) = N_c(0) + [N_o(0)\lambda_{ic}N_i(0) - \lambda_{ci}N_c(0)]\Delta t$$

This yields 9 new values for the 9 functions, all evaluated at t equal Δt.

These 9 values are the initial values for the next time step, from Δt to $2\Delta t$. By analogy with Equation 31:

(6.41) $$N(\Delta t+\Delta t) = N(2\Delta t) = N(\Delta t) + [\Delta N(\Delta t)/\Delta t]\Delta t$$

where $\Delta N(\Delta t)/\Delta t$ is calculated from the right-hand sides of Equations 6.13 through 6.21, with N(t) replaced by $N(\Delta t)$. In other words, it is assumed that the slope of N(t) over the interval Δt to $2\Delta t$ equals the slope of N(t) evaluated at Δt. The analogues of Equations 6.21 through 6.40 are:

(6.42) $$N_s(2\Delta t) = N_s(\Delta t) + [R_{in} + \lambda_{ws}N_w(\Delta t) - \lambda_{sw}N_s(\Delta t)]\Delta t$$

(6.43) $$N_w(2\Delta t) = N_w(\Delta t) + [\lambda_{sw}N_s(\Delta t) + \lambda_{pw}N_p(\Delta t) - \lambda_{ws}N_w(\Delta t) - \lambda_{wp}N_w(\Delta t)]\Delta t$$

(6.44) $$N_p(2\Delta t) = N_p(\Delta t) + [\lambda_{wp}N_w(\Delta t) - \lambda_{pw}N_p(\Delta t)]\Delta t$$

(6.45) $$N_{GI}(2\Delta t) = N_{GI}(\Delta t) + [N_s(\Delta t)IR_s(\Delta t)/V_s + N_w(\Delta t)IR_w(\Delta t)/V_w + N_p(\Delta t)IR_p(\Delta t)/V_p - \lambda_{GIb}N_{GI}(\Delta t)]\Delta t$$

(6.46) $$N_b(2\Delta t) = N_b(\Delta t) + [\lambda_{GIb}N_{GI}(\Delta t) + \lambda_{ob}N_o(\Delta t) - \lambda_{b,out}N_b(\Delta t) - \lambda_{bo}N_b(\Delta t)]\Delta t$$

(6.47) $$N_o(2\Delta t) = N_o(\Delta t) + [\lambda_{bo}N_b(\Delta t) - \lambda_{ob}N_o(\Delta t)]\Delta t$$

(6.48) $$N_n(2\Delta t) = N_n(\Delta t) + [\lambda_{in}N_i(\Delta t) - N_o(\Delta t)\lambda_{ni}N_n(\Delta t)]\Delta t$$

(6.49) $$N_i(2\Delta t) = N_i(\Delta t) + [N_o(\Delta t)\lambda_{ni}N_n(\Delta t) + \lambda_{ci}N_c(\Delta t) - \lambda_{in}N_i(\Delta t) - N_o(\Delta t)\lambda_{ic}N_i(\Delta t)]\Delta t$$

(6.50) $$N_c(2\Delta t) = N_c(\Delta t) + [N_o(\Delta t)\lambda_{ic}N_i(\Delta t) - \lambda_{ci}N_c(\Delta t)]\Delta t$$

This process of time-steps continues until the entire interval over which the differential equation is being solved has been integrated. The general process may be summarized as:

(6.51) $$N(n\Delta t+\Delta t) = N(n\Delta t) + [\Delta N(n\Delta t)/\Delta t]\Delta t$$

or

(6.51) $$N(n\Delta t) = N(n\Delta t-\Delta t) + [\Delta N(n\Delta t-\Delta t)/\Delta t]\Delta t$$

where n is the number of time-steps that have taken place. Again, the approximation to $N(n\Delta t)$ will improve as Δt is made smaller (which means the value of n corresponding to the upper limit of the integration increases).

As with numerical integration (see Section 6.2), the size of the time-step, Δt, must be sufficiently small to allow reasonable approximations. The required size depends on the second and higher derivatives of N(t). Where these derivatives are large, the slope will not be constant over large increments Δt and the approximations introduced by Euler's method will not be reliable. To determine if the approximation is reliable, the same method adopted to choose a time-step for numerical integration may be applied:

1. Establish a desired maximal allowed change in the solution calculated between any two selections of the time step Δt, (e.g. 0.01 or 1%).

2. Select an initial time-step, Δt.

3. Obtain the solution (i.e. prediction of N(t)) to the differential equation with this time-step. This solution is $S_{\Delta t}$.

4. Divide Δt by 2.

5. Calculate the solution using time-steps of $\Delta t/2$. This new solution is $S_{\Delta t/2}$.

6. Calculate $(S_{\Delta t}-S_{\Delta t/2})/S_{\Delta t}$.

7. If this ratio is greater than the maximal allowed change from Step 1 (e.g. greater than 0.01), return to Step 4 and decrease Δt; otherwise, accept this new solution.

Also in analogy to the process of numerical integration, the solution to a differential equation will be less efficient if Δt is the same over the entire interval of solution. Where the second derivative is large, Δt must be small; where the second derivative is small, Δt may be large without loss of accuracy. Only methods with a constant value of Δt over the interval of integration are considered here.

Euler's method as applied to a system of differential equations may also be formulated at times as a matrix problem. Let N(t) be a column matrix (i.e. one column and n rows) of the solutions to the differential equations defining a system. For this example, consider a system of two differential equations of the form:

$$(6.52)\quad dN_1(t)/dt = -\lambda_{12}N_1(t) + \lambda_{21}N_2(t) - \lambda_{1,out}N_1(t)$$

$$(6.53)\quad dN_2(t)/dt = \lambda_{12}N_1(t) - \lambda_{21}N_2(t) - \lambda_{2,out}N_2(t)$$

where λ_{12} is the first order rate constant for transfers between compartments 1 and 2; λ_{21} is the first order rate constant for transfers between compartments 2 and 1; $\lambda_{1,out}$ is the first order rate constant for transfers from compartment 1 to outside the system; and $\lambda_{2,out}$ is the first order rate constant for transfers from compartment 2 to outside the system. Define a matrix L consisting of transfer terms:

$$L = \begin{vmatrix} -\lambda_{12}-\lambda_{1,out} & \lambda_{21} \\ \lambda_{12} & -\lambda_{21}-\lambda_{2,out} \end{vmatrix}$$

Equations 6.52 and 6.53 may then be written as the matrix equation:

$$(6.54)\quad \begin{vmatrix} dN_1(t)/dt \\ dN_2(t)/dt \end{vmatrix} = \begin{vmatrix} -\lambda_{12}-\lambda_{1,out} & \lambda_{21} \\ \lambda_{12} & -\lambda_{21}-\lambda_{2,out} \end{vmatrix} \times \begin{vmatrix} N_1(t) \\ N_2(t) \end{vmatrix}$$

Further, the numerical solution may be written as:

$$(6.55)\quad \begin{vmatrix} N_1(n\Delta t+\Delta t) \\ N_2(n\Delta t+\Delta t) \end{vmatrix} = \begin{vmatrix} N_1(n\Delta t) \\ N_2(n\Delta t) \end{vmatrix} + \begin{vmatrix} -\lambda_{12}-\lambda_{1,out} & \lambda_{21} \\ \lambda_{12} & -\lambda_{21}-\lambda_{2,out} \end{vmatrix} \times \begin{vmatrix} N_1(n\Delta t) \\ N_2(n\Delta t) \end{vmatrix}$$

The numerical solution may be generated through a series of matrix operations.

Example 6.2. Consider the following system of differential equations:

$$dN_1(t)/dt = -0.1N_1(t) + 0.3N_2(t)$$
$$dN_2(t)/dt = 0.1N_1(t) - 0.3N_2(t)$$

with initial values for $N_1(0)$ and $N_2(0)$ of 30 and 50 grams, respectively. The transfer rate constants have units of day^{-1}. The time-step is 0.1 day. The solution to both functions at t equal 0.2 days is desired. After 0.1 day, the solutions are:

$$N_1(0.1) = N_1(0)+[-0.1N_1(0) + 0.3N_2(0)]\Delta t = 30 + [-0.1\bullet 30 + 0.3\bullet 50]\bullet 0.1 = 31.2\text{ g}$$

$$N_2(0.1) = N_2(0) + [0.1N_1(0) - 0.3N_2(0)]\Delta t = 50 + [0.1\bullet 30 - 0.3\bullet 50]\bullet 0.1 = 48.8\text{ g}$$

and after 0.2 day:

$$N_1(0.2) = N_1(0.1) + [-0.1N_1(0.1) + 0.3N_2(0.1)]\Delta t$$

$$= 31.2 + [-0.1\bullet 31.2 + 0.3\bullet 48.8]\bullet 0.1 = 32.35\text{ grams}$$

$$N_2(0.2) = N_2(0.1) + [0.1N_1(0.1) - 0.3N_2(0.1)]\Delta t$$

$$= 48.8 + [0.1\bullet 31.2 - 0.3\bullet 48.8]\bullet 0.1 = 47.65\text{ grams}$$

6.4. Numerical Solutions to Differential Equations: Runge-Kutta Methods

Euler's method for solving ordinary differential equations is the most conceptually easy, and the most computationally simple, but it is also the least accurate method for a given size of time-step. The choice of time-step may need to be quite small for that method, since the error introduced by the approximation is only one power lower than the correction (i.e. the slope). The reason is that only information on the slope obtained at the beginning of a time-step interval is used.

Better methods, and the ones used most commonly in environmental risk assessment, are a class of methods known as *Runge-Kutta* [6]. They make use of estimates of the slope of the desired function at points in the time-step interval other than only at the beginning of the interval. In the Runge-Kutta 2 method, the derivative on the right-hand side of the differential equation is evaluated twice, once at the beginning of the interval and once at the midpoint of the interval. In the Runge-Kutta 4 method, the derivative is evaluated 4 times for each time-step. In general, the Runge-Kutta 4 method is the most accurate of the three methods discussed here, although at the sacrifice of a greater number of calculations to reach the end of the integration (essentially twice the number of calculations required by Runge-Kutta 2 and four times the number of calculations required by Euler's method for the same step size, with a resulting increase in computational time to complete the integration over an interval).

Equation 6.51 may be examined to better understand the Runge-Kutta methods:

$$N(n\Delta t+\Delta t) = N(n\Delta t) + [\Delta N(n\Delta t)/\Delta t]\Delta t \tag{6.56}$$

Consider the slope of N(t) evaluated at the beginning of a time-step, $n\Delta t$. We can define the correction (or amount to be added during the time step) to be applied to $N(n\Delta t)$ as:

$$k_1 = [\Delta N(n\Delta t)/\Delta t]\Delta t \tag{6.57}$$

where $\Delta N(n\Delta t)/\Delta t$ is evaluated using $N(n\Delta t)$ wherever N(t) appears on the right-hand side of the differential equation. Then consider evaluating the slope of N(t) at the middle of the time-step, $n\Delta t + \Delta t/2$. This would produce a correction to be applied to $N(n\Delta t)$ as:

$$k_2 = [\Delta N(n\Delta t+\Delta t/2)/\Delta t]\Delta t \tag{6.58}$$

where $\Delta N(n\Delta t+\Delta t/2)/\Delta t$ is evaluated using $N(n\Delta t)+k_1/2$ wherever N(t) appears on the right-hand side of the differential equation. Note that k_1 is calculated as in Euler's method to estimate an initial slope at the beginning of the time-step interval. This slope then is followed to the middle of the interval and new values of N(t) and the slope of N(t) calculated here. This new slope then is applied to the entire interval from $n\Delta t$ to $n\Delta t + \Delta t$. The resulting process may be written as:

$$N(n\Delta t+\Delta t) = N(n\Delta t) + k_2 \tag{6.59}$$

The error in this term is second order, which is an improvement over the error term in Euler's method. This means that, for a given time-step size, Runge-Kutta 2 will produce a more accurate estimate of the solution, although at the expense of twice the number of calculations and a greater computation time.

Example 6.3. Consider a single differential equation:

$$dN(t)/dt = -0.3N(t)$$

It may be noted from Chapter 3 that this is a simple first-order equation with solution:

$$N(t) = N(0)e^{-0.3t}$$

N(0) is 10 grams. Let the time-step be 0.1 (in days if 0.3 is in day^{-1}). Using Euler's method:

$$N(0.1) = N(0) + [-0.3N(0)]\bullet 0.1 = 10 + [-0.3\bullet 10]\bullet 0.1 = 9.7 \text{ grams}$$

Using Runge-Kutta 2:

$$k_1 = [-0.3N(0)]\bullet 0.1 = -0.3\bullet 10\bullet 0.1 = -0.3$$

$$k_2 = [-0.3\{N(0)+k_1/2\}]\bullet 0.1 = [-0.3\{10-0.3/2\}]\bullet 0.1 = -0.2955$$

$$N(0.1) = N(0) + k_2 = 10 - 0.2955 = 9.7045 \text{ grams}$$

From the analytic solution, $N(0.1) = 10e^{-0.3\bullet 0.1} = 9.70445$ grams

Note the improvement in accuracy through the Runge-Kutta 2 method (but count the number of computations required by each method).

Even greater accuracy may be obtained with the Runge-Kutta 4 method, although with a greater number of computational steps needed to reach the end of a given time-step interval. Five separate calculations are performed; four are calculations of slope and values throughout the interval and the fifth combines this information with the value of the function at the beginning of the interval to yield the approximation at the end of the interval. Consider the slope of N(t) evaluated at the beginning of a time-step, $n\Delta t$. We can define the correction to be applied to $N(n\Delta t)$ as:

$$k_1 = [\Delta N(n\Delta t)/\Delta t]\Delta t \tag{6.60}$$

where $\Delta N(n\Delta t)/\Delta t$ is evaluated using $N(n\Delta t)$ wherever N(t) appears on the right-hand side of the differential equation. Then consider evaluating the slope of N(t) at the middle of the time-step, $n\Delta t + \Delta t/2$. This would produce a correction to be applied to $N(n\Delta t)$ as:

$$k_2 = [\Delta N(n\Delta t+\Delta t/2)/\Delta t]\Delta t \tag{6.61}$$

where $\Delta N(n\Delta t)/\Delta t$ is evaluated using $N(n\Delta t)+k_1/2$ wherever N(t) appears on the right-hand side of the differential equation.

Then consider evaluating the slope of N(t) at the middle of the time-step, $n\Delta t + \Delta t/2$, yet again. This would produce a correction to be applied to $N(n\Delta t)$ as:

$$k_3 = [\Delta N(n\Delta t+\Delta t/2)/\Delta t]\Delta t \tag{6.62}$$

where $\Delta N(n\Delta t)/\Delta t$ is evaluated using $N(n\Delta t)+k_2/2$ wherever N(t) appears on the right-hand side of the differential equation. Finally, evaluate the slope of N(t) at the end of the time-step, $n\Delta t+\Delta t$. This would produce a correction to be applied to $N(n\Delta t)$ as:

$$k_4 = [\Delta N(n\Delta t+\Delta t)/\Delta t]\Delta t \tag{6.63}$$

where $\Delta N(n\Delta t)/\Delta t$ is evaluated using $N(n\Delta t)+k_3$ wherever N(t) appears on the right-hand side of the differential equation. The resulting approximation to $N(n\Delta t+\Delta t)$ may be written as:

$$N(n\Delta t+\Delta t) = N(n\Delta t) + k_1/6 + k_2/3 + k_3/3 + k_4/6 \tag{6.64}$$

The error in this term is fourth order, which is an improvement over the error term in both of the previous methods, although at the expense of several more calculations for each time-step interval. This improvement can be significant when dealing with complex systems of coupled differential equations that are highly non-linear. It also may be an important improvement if the numerical values of rate constants in the equations span several orders of magnitude.

Example 6.4. Consider the same differential equation as in Example 6.3:

$dN(t)/dt = -0.3N(t)$

Again, N(0) is 10 grams and the time-step is 0.1 day. Using Runge-Kutta 4:

$k_1 = [-0.3N(0)]\bullet 0.1 = -0.3\bullet 10\bullet 0.1 = -0.3$

$k_2 = [-0.3\{N(0)+k_1/2\}]\bullet 0.1 = [-0.3\{10-0.3/2\}]\bullet 0.1 = -0.2955$

$k_3 = [-0.3\{N(0)+k_2/2\}]\bullet 0.1 = [-0.3\{10-0.2955/2\}]\bullet 0.1 = -0.2955675$

$k_4 = [-0.3\{N(0)+k_3/2\}]\bullet 0.1 = [-0.3\{10-0.2955675\}]\bullet 0.1 = -0.291133$

Therefore:

$N(0.1) = N(0) + k_1/6 + k_2/3 + k_3/3 + k_4/6$

$= 10 - 0.3/6 - 0.2955/3 - 0.2955675/3 - 0.291133/6 = 9.7044$

As with Euler's method, the Runge-Kutta methods require that the numerical solutions proceed in a distinct order when solving coupled systems of differential equations. For any given step (e.g. calculation of k_1 for the first time-step), the step must be performed for all differential equations in the system before proceeding to the next step for any of the differential equations. The reason is that each step requires information on the results of the previous steps from several of the differential equations (assuming the equations are coupled, which is the case for the problems considered in this chapter).

6.5. The STELLA Modeling Software

As the use of numerical methods has grown, a variety of software packages have been created to assist in the development of models. One of the most popular due to the ease of use is the STELLA package created by High Performance Systems (see their web site at www.hps-inc.com). This package is considered here simply as an example of the use of such software products in generating numerical solutions to models. The software has the advantage of being relatively inexpensive and conceptually simple to grasp. It serves, therefore, as an excellent starting point for learning how to build models requiring numerical solutions.

As with all such modeling systems, STELLA is provided with a graphical user interface (GUI) to organize the tasks. A workspace (essentially a blank slate or page) is provided on which to construct the model. The following tools then are provided; each may be dragged onto the workspace to create a model:

- *Compartments*, shown in this package as a box (see Figure 6.7). These represent the compartments of the system being modeled.

- *Pipes/Valves*, shown in this package as a pipe with a valve in the center. These represent flows between compartments, or flows into and out of the system.

- *Regulators*, shown in this package as a circle. These represent numerical values that control valves (e.g. transfer rate constants).

- *Connectors*, shown in this package as arrows. These represent functional dependencies between Compartments, Valves and Regulators/

As an example of a model that might be constructed within STELLA, consider the state-vector model of carcinogenesis in Figure 6.3. The defining differential equations are repeated here:

$$dN_n(t)/dt = \lambda_{in}N_i(t) - N_o(t)\lambda_{ni}N_n(t) \quad (6.9)$$

$$dN_i(t)/dt = N_o(t)\lambda_{ni}N_n(t) + \lambda_{ci}N_c(t) - \lambda_{in}N_i(t) - N_o(t)\lambda_{ic}N_i(t) \quad (6.10)$$

$$dN_c(t)/dt = N_o(t)\lambda_{ic}N_i(t) - \lambda_{ci}N_c(t) \quad (6.11)$$

Since STELLA does not display subscripts, the equations are formatted slightly differently below so the terms will appear in the same format in the text and the subsequent figures:

$$dNn(t)/dt = Lin\ Ni(t) - No(t)\ Lni\ Nn(t) \quad (6.65)$$

$$dNi(t)/dt = No(t)\ Lni\ Nn(t) + Lci\ Nc(t) - Lin\ Ni(t) - No(t)\ Lic\ Ni(t) \quad (6.66)$$

(6.67) $$dNc(t)/dt = No(t)\ Lic\ Ni(t) - Lci\ Nc(t)$$

Figure 6.7 shows the STELLA model created for this particular series of differential equations.

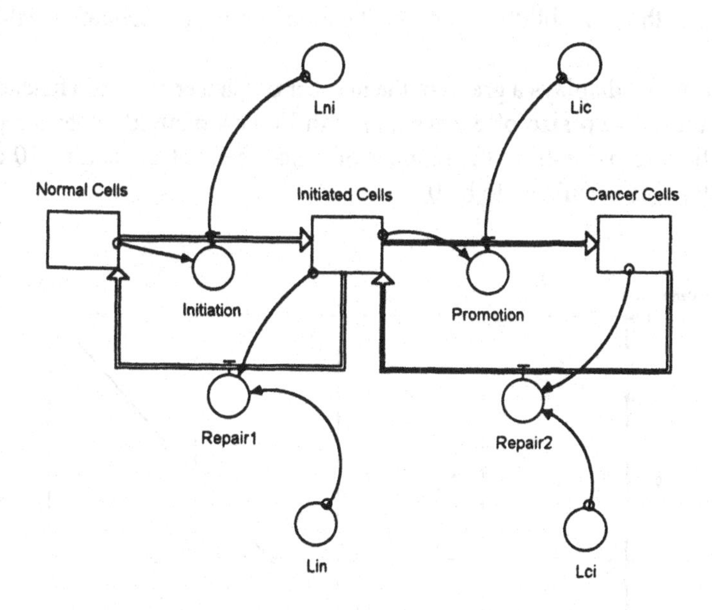

Figure 6.7. The STELLA model version of Equations 6.65 through 6.67. The boxes are Compartments; the circles are Regulators; the Pipes and Valves show flow between the Compartments; the arrows are Connectors.

In Figure 6.7, the mathematical relationships for the Valves are the same as those appearing on the right-hand sides of Equations 6.65 through 6.67. Specifically, the Valve shown as Initiation in the Figure is "open" by an amount equal to the product of Lni and the number of Normal Cells. The Valve shown as Promotion in the Figure is "open" by an amount equal to the product of Lic and the number of Initiated Cells. The Valve shown as Repair1 in the Figure is "open" by an amount equal to the product of Lin and the number of Initiated Cells. The Valve shown as Repair2 in the Figure is "open" by an amount equal to the product of Lci and the number of Cancer Cells. The initial values of the number of cells in each of the three Compartments or States have also been specified. In the following example, they are 1 for Nn(0); 0 for Ni(0) and 0 for Nc(0). In other words, all cells are Normal prior to any exposure to the pollutant. In the same example, Lni is 0.1 per day; Lic is 0.5 per day; Lci is 0.03 per day; and Lin is 1 per day (*Note*: these values are provided here only for purpose of an example calculation. They do not represent realistic values for any particular pollutant).

STELLA allows plotting of any of the quantities shown in Figure 6.7 as a function of time, using tables and/or graphs. For this example, only the number of cells in the Cancer state will be calculated, since this will be considered proportional to the probability of cancer. The time interval of integration will be 0 to 10 days.

Figure 6.8 displays a graph of the number of Cancer cells as a function of time using a time step size of 5 days and with Euler's method. Note the poor quality of the approximation; the number of cancer cells at the end of 10 days (two time steps) is calculated to be 0.5.

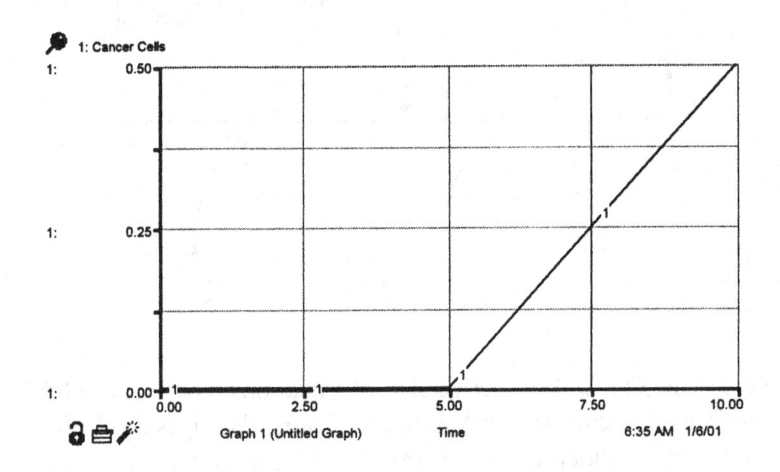

Figure 6.8. The result of the numerical simulation using STELLA with a time-step size of 5 days, an integration interval of 10 days, and Euler's method. Note that there is a prediction of no Cancer cells up until the end of the first time step, since the slope of the number of Cancer cells is 0 at the start of the time step.

Figure 6.9 displays the same modeling result, with the same conditions, but with Runge-Kutta 2. Note that the number of Cancer cells increases even during the first time-step, since the slope in the middle of the time-step is used rather than only the slope at the beginning of the time-step. Note also that the prediction of the number of Cancer cells at the end of 10 days is less than 0.5.

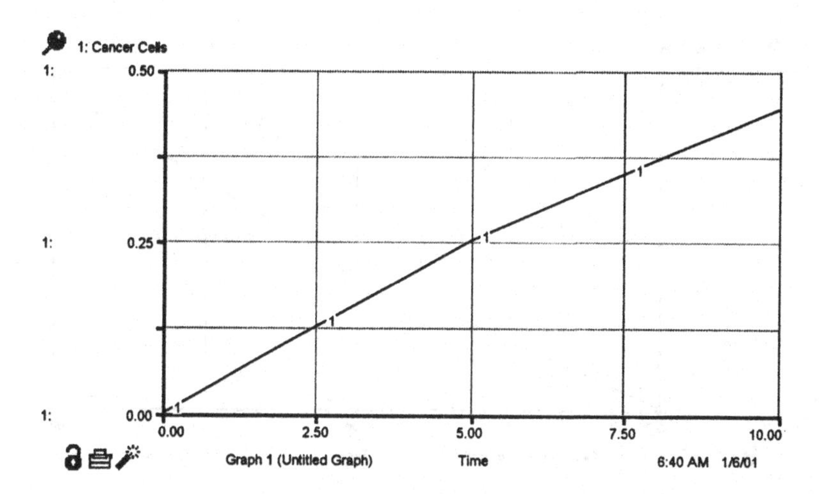

Figure 6.9. The result of the numerical simulation using STELLA with a time-step size of 5 days, an integration interval of 10 days, and the Runge-Kutta 2 method. Note that there is a prediction of increased numbers of Cancer cells during the first time step (compare to Figure 6.8).

Figure 6.10 displays the same modeling result, with the same conditions, but with Runge-Kutta 4. Note again that the number of Cancer cells increases even during the first time-step, since the slope at several points during time-step is used rather than only the slope at the beginning of the time-step. Note also that the prediction of the number of Cancer cells at the end of 10 days is less than 0.5, and that the shape of the curve differs from either the Euler's or Runge-Kutta 2 curves.

Figures 6.11 through 6.13 display the results of the model using Euler's method but with increasingly small time-step sizes. Note the progressive improvement of the curve, with little difference between the results in Figures 6.12 and 6.13. A time-step of 0.1 is sufficient to obtain reasonable accuracy (with the answer at 10 days being within 1% of the correct answer). In general, the maximum acceptable time-step may be found from the largest rate constant. In this problem, the largest rate constant is 1 per day. Take the inverse of this largest rate constant (here, that inverse is 1). The maximum time-step should not be larger than approximately 10% of this value. As a result, a time-step of 1/10 or 0.1 day would be acceptable.

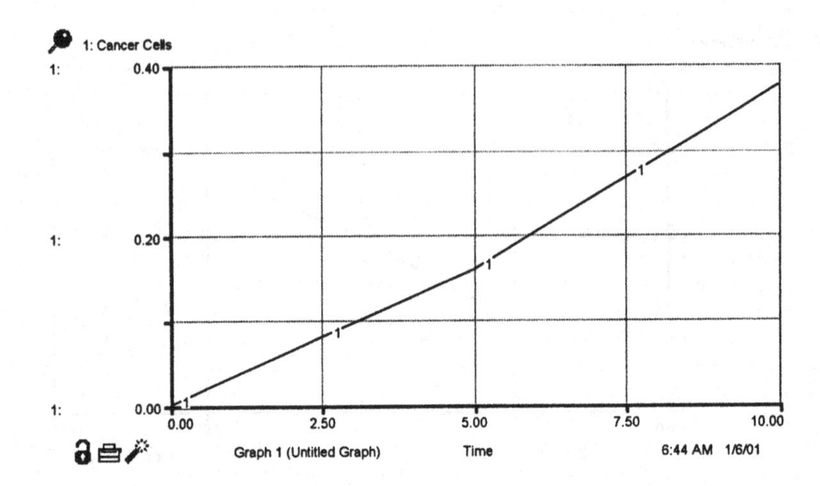

Figure 6.10. The result of the numerical simulation using STELLA with a time-step size of 5 days, an integration interval of 10 days, and the Runge-Kutta 4 method. Note that there is a prediction of increased numbers of Cancer cells during the first time step (compare to Figure 6.8).

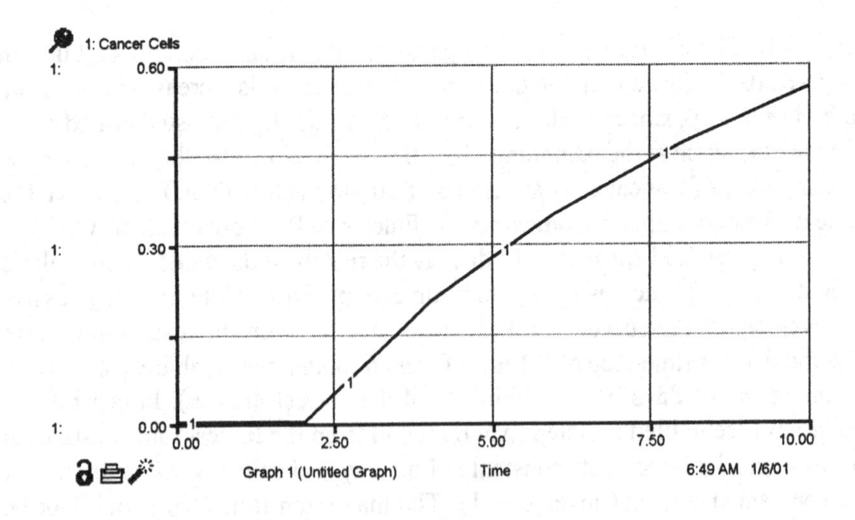

Figure 6.11. The result of the numerical simulation using STELLA with a time-step size of 2 days, an integration interval of 10 days, and Euler's method.

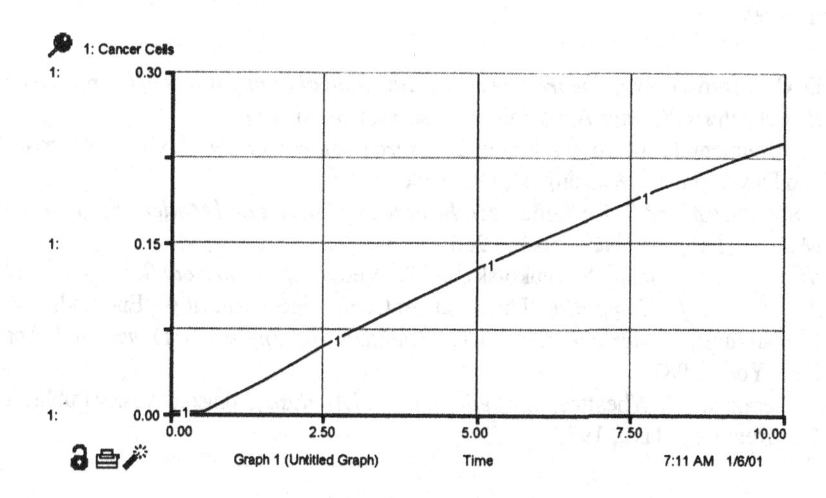

Figure 6.12. The result of the numerical simulation using STELLA with a time-step size of 0.5 days, an integration interval of 10 days, and Euler's method.

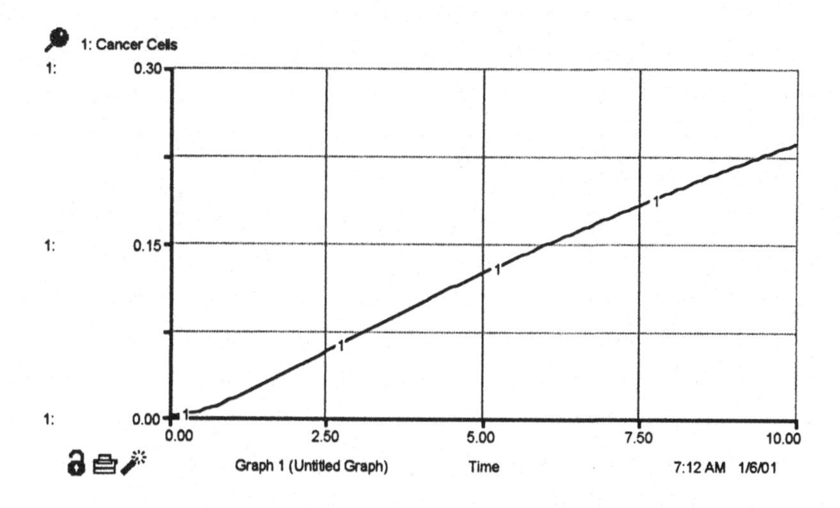

Figure 6.13. The result of the numerical simulation using STELLA with a time-step size of 0.1 days, an integration interval of 10 days, and Euler's method.

References

1. D. Crawford-Brown, *Theoretical and Mathematical Foundations of Human Health Risk Analysis*, Kluwer Academic Publishers, Boston, 1997.
2. Environmental Protection Agency, *Exposure Factors Handbook*, Office of Research and Development, Washington, DC, 1996.
3. I. Sokolnikoff and R. Redheffer, *Mathematics of Physics and Modern Engineering*, McGraw-Hill, Inc., New York, 1966.
4. W. Press, B. Flannery, S. Teukolsky and W. Vetterling, *Numerical Recipes in C: The Art of Scientific Computing*, University of Cambridge, Cambridge, England, 1988.
5. R. Hamming, *Numerical Methods for Scientists and Engineers*, Dover Publishers, New York, 1987.
6. C. Gerald and P. Wheatley, *Applied Numerical Analysis*, Addison-Wesley Publishing Co., Reading, Mass., 1998.

CHAPTER 7
Monte Carlo Methods

7.1. Decisions Under Variability and Uncertainty

The risk to individuals in a population is clouded by several issues that must be accounted for in risk-based decisions. These issues usually are treated in the risk characterization stage of a risk assessment:

- Different individuals will experience different risks due to variability in the factors that control exposure, susceptibility and sensitivity. These differences arise from variation in location within the exposure field, variation in exposure factors (such as breathing rates or ingestion rates), variation in pharmacokinetic and pharmacodynamic properties, and variation in dose-response characteristics. The composite effect of these differences is that the risk to individuals in a population must be described by an *intersubject variability probability density function* rather than by a single estimate (often called a *point estimate* in risk assessment).

- Neither the parameters appearing in models, nor the mathematical forms of the model, are known with complete accuracy. As a result, there is uncertainty in predictions of risk, either for an individual or population. This requires that risk estimates be characterized by an *uncertainty probability density function* describing the confidence associated with which each value of the risk.

- Some processes appear to be *stochastic*, with random behavior at their most fundamental level. For example, the interaction of radiation with the DNA of cells, or the length of time a specific molecule remains in the target organ, may be stochastic processes. The probability of effect in such cases is a function of this stochastic behavior and must be accounted for in the most complete scientific models of these processes. Again, the result is uncertainty in risk estimates The source of the uncertainty differs from that discussed in the second bullet, however, since in the former

> case the uncertainty was caused by problems of model formulation and parameter measurement (and so the uncertainty can be reduced through better analyses), whereas the uncertainty in the present case is an irreducible feature of the reality being modeled.

Risk assessments rarely make use of stochastic models. The reason is that individuals usually are exposed to many molecules and/or quanta of radiation (on the order of millions in a typical exposure), and so the stochastic behavior is averaged out. The other two sources of variability and uncertainty do, however, play a central role in modern risk assessments.

It is recognized that it may not be possible to protect 100% of the population with 100% confidence. Instead, a compromise is reached between such an ideal goal and what is practical. As a result, the goal of a risk-based policy can be stated as:

Find a policy (e.g. control option on an industrial facility) that will ensure with at least X% confidence (e.g. 95% confidence) that no more than Y% of the population (e.g. 1%) will have a risk exceeding Z (e.g. 10^{-4}).

Locating such a policy requires generating intersubject variability and uncertainty PDFs for the estimates of risk in the population. This chapter reviews some of the basic methods by which such PDFs may be produced using first analytic methods and then Monte Carlo methods. While the same Monte Carlo methods may be applied to problems of modeling stochastic processes, such applications are not considered here.

7.2. Analytical Methods

In some relatively simple cases, it is possible to calculate the characteristics of intersubject variability and/or the uncertainty PDFs associated with estimates of risk directly from the PDFs for the underlying parameters. This will be true only if uncertainty in model formulation is ignored, but ignoring this source of uncertainty is common in risk assessment (we will consider later how to incorporate it into the assessment). Some examples were provided in Chapter 2, Section 2.8, based on the formulation of variance shown in Equation 2.25.

In this section, an analytic approach to intersubject variability for multiplicative models when parameters are distributed lognormally is considered. Such models will arise when, for example, dispersion models are reduced to dispersion coefficients, when systems models are reduced to equilibrium ratios, and the risk is assumed to be proportional to concentration in the environment.

For example, consider the case in which the source term for a pollutant from a facility is ST (g/day), the dispersion coefficient in the atmosphere is DC (g/m^3 per g/day), the equilibrium ratio of soil concentration over air concentration is $ER_{s/a}$, and the slope factor for the probability of effect (e.g. cancer) per unit concentration of the pollutant in soil is SF (m^3/g). The probability of cancer may then be expressed as:

$$P = ST \bullet DC \bullet ER_{s/a} \bullet SF \tag{7.1}$$

Note that the probability is a multiplicative function of the four parameters.

Consider next the uncertainty and/or intersubject variability in P. If intersubject variability is being considered, the goal is to determine the best estimate of the PDF describing the variation in the values of P across a population. For each of the four parameters in Equation 7.1, it is necessary to specify the best estimate of the variability of the parameter value across individuals in the population. If uncertainty is being considered, the goal is to determine the PDF describing the uncertainty in the mean value of P for the population. For each of the four parameters in Equation 7.1, it is necessary to specify the uncertainty in the mean parameter value for the population. The same mathematical technique applies to uncertainty and variability analysis, and so only variability analysis will be considered further in this section.

Let the intersubject variability in each of the four parameters be described by a lognormal PDF (see Equation 2.9). As described in Chapter 2, such PDFs are characterized by a *median* or *geometric mean* (two names for the same quantity) and a geometric standard deviation or GSD. The median is the value of the parameter for which 50% of individuals have a smaller value and 50% have a larger value. 68% of individuals have a value of the parameter lying in the interval [Median/GSD, Median•GSD], with 34% lying in the interval [Median/GSD, Median] and 34% lying in the interval [Median, Median•GSD]. Note that the ratio of the 84th percentile over the 50th percentile, and the ratio of the 50th percentile over the 16th percentile, equals the GSD.

Let Med_i be the median value of the ith parameter in a population of people and GSD_i be the GSD of that same parameter. If all parameters are distributed lognormally, if there is no truncation of the parameters, if there is no correlation between parameters, and if the model in which they appear is multiplicative (as is the case in Equation 7.1), then the following relationships hold [1]:

$$Med_P = \Pi\, Med_i \tag{7.2}$$

where Π indicates the product of the individual median values, taken over all values of i from 1 to n, and Med_P is the median of the product of the parameter values. Also [1]:

$$(7.3) \quad GSD_P = \exp[LN^2(GSD_1) + LN^2(GSD_2) + \ldots LN^2(GSD_n)]^{0.5}$$

where GSD_P is the GSD of the product of the parameter values and GSD_i is the GSD of the ith parameter. Note that the sum in the exponent on the right-hand side is from i equal 1 to n for n parameters. Applying Equations 7.2 and 7.3 to the case of Equation 7.1 yields:

$$(7.4) \quad Med_P = Med_{ST} \bullet Med_{DC} \bullet Med_{ER} \bullet Med_{SF}$$

and

$$(7.5) \quad GSD_P = \exp[LN^2(GSD_{ST}) + LN^2(GSD_{DC}) + LN^2(GSD_{ER}) + LN^2(GSD_{SF})]^{0.5}$$

Example 7.1. For Equation 7.1, let the median for the intersubject variability distribution describing ST be 2 and the GSD be 3; let the median for the intersubject variability distribution describing DC be 0.1 and the GSD be 2; let the median for the intersubject variability distribution describing ER be 4 and the GSD be 2.5; and let the median for the intersubject variability distribution describing SF be 0.0001 and the GSD be 4. The median for the probability of cancer in the population, using Equation 7.2 or 7.4, will be:

$$P = Med_{ST} \bullet Med_{DC} \bullet Med_{ER} \bullet Med_{SF} = 2 \bullet 0.1 \bullet 4 \bullet 0.0001 = 0.00008$$

and the GSD (using Equation 7.3 or 7.5) is:

$$GSD_P = \exp[LN^2(GSD_{ST}) + LN^2(GSD_{DC}) + LN^2(GSD_{ER}) + LN^2(GSD_{SF})]^{0.5}$$

$$= \exp[LN^2(3) + LN^2(2) + LN^2(2.5) + LN^2(4)]^{0.5} = 8.24$$

The intersubject variability in P then is characterized by a lognormal distribution PDF with a median of 0.00008 and a GSD of 8.24. The 68% confidence interval is [0.00008/8.24, 0.00008 • 8.24] = 0.0000098, 0.00066].

A useful feature of lognormal distributions is that their cumulative distribution functions appear as straight lines on log probit paper [2]. As a result, it is necessary to know only two points in generating the complete CDF. For example, if the median and GSD are known (see Example 7.1), the CDF can be plotted on log probit axes using the two points of the median and the 84th percentile (median times the GSD) or the 16th percentile (median divided by the GSD). Conversely, if the CDF is available on the log probit axes (perhaps obtained by fitting a straight line to quantiles), the GSD can be calculated by dividing the 84th percentile value by the 50th percentile value of the CDF, or the 50th percentile value by the 16th percentile value.

7.3. Monte Carlo Methods

Only the simplest risk assessment models are susceptible to variability and uncertainty analysis using analytic methods. In more complex cases, generation of the PDF or CDF for the model predictions requires different techniques. By far the most common technique used in risk assessment is Monte Carlo analysis based on random sampling from the PDFs characterizing the individual parameters in a model [3].

Figure 7.1 displays the logic tree associated with Equation 7.1 (left branch of the tree). In addition, a branch has been added using a quadratic relationship between soil concentration and probability of effect (right branch), indicating uncertainty in the form of the exposure-response model. Note that the premises are of two general types: premises related to the choice of model (Premises 3, 5, 6 and 8) and premises related to parameter values used in the model (all other premises). Only the left branch, with a linear relationship, is used in this section. Consideration of uncertainty in model form will be introduced in Section 7.4. As a result, the uncertainty or variability analysis considered here is a *conditional analysis* (i.e. conditional upon the particular linear model appearing in the left branch of the tree). The following discussion focuses on variability analysis as in Section 7.2, although again the same methods may be applied to uncertainty analysis (and will be applied in Section 7.4).

For this conditional analysis, Monte Carlo methods begin by replacing the premises related to parameter values (Premises 1, 2, 4 and 7) by the probability density functions characterizing intersubject variability in these parameters. The task is to select a value at random from each of these four PDFs (we will assume independence between parameters), form the quadruplet (SF, DC, ER, SF), place this quadruplet into Equation 7.1, and obtain a single estimate of P (called a single *realization* or *simulation* of P). This process is repeated many times, with a range of values of P calculated, and the results summarized as a PDF characterizing intersubject variability in P for the exposed population.

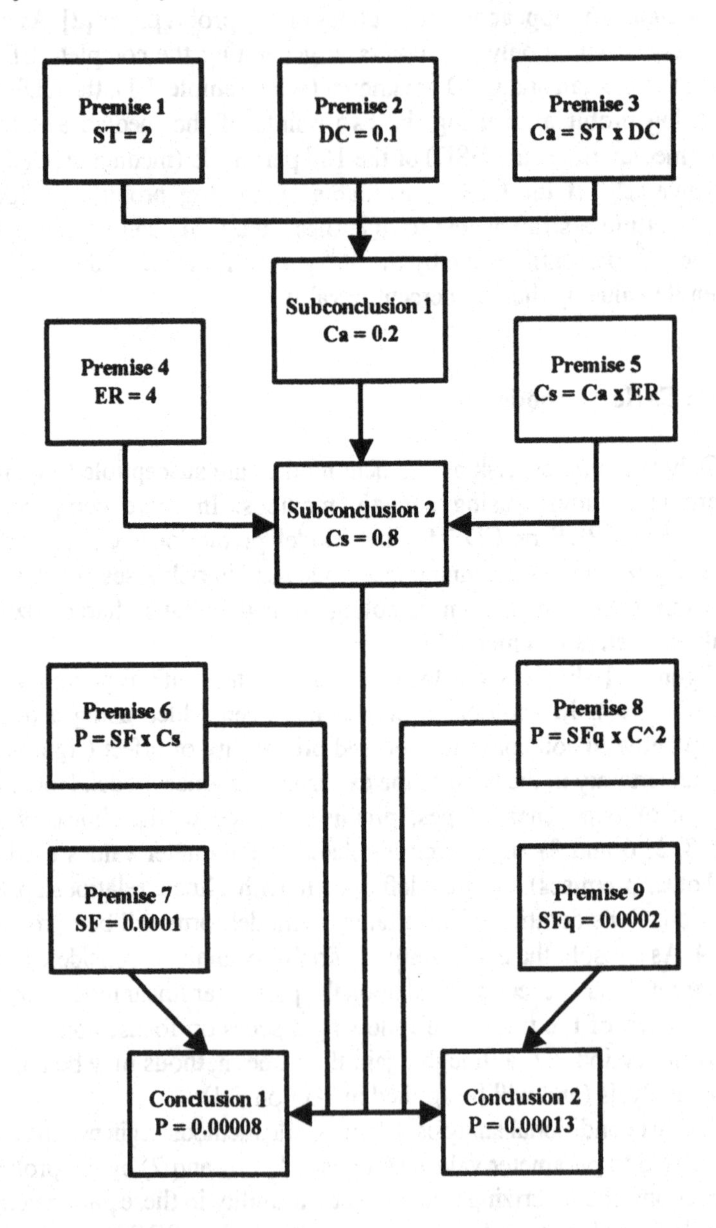

Figure 7.1. An example logic tree with uncertainty as to model form for the exposure-response relationship. The numerical values of the premises are the median values discussed in this section. Ca is the concentration in air, Cs is the concentration in soil, and P is the probability of effect.

How are the values for each parameter in the quadruplet to be selected at random? Consider a PDF for a random variable defined in the interval [0,1] and with a uniform PDF (see Figure 7.2). Under these conditions, all values of the variable will have an equal PDF. If selections are made from this PDF, it is equally likely that any value in the interval from 0 to 1 will be selected. For example, it is equally likely that 0.1, 0.4, 0.713, 0.21774, etc, will be selected. Further, if we ask for the probability that a value equal to *X or less* will be selected from this uniform PDF, the answer is exactly X. The reason can be seen by noting that the cumulative distribution function corresponding to this PDF is a straight line whose magnitude at any value of X is simply X (see Figure 7.3; note that the CDF is not defined above 1). The probability of selecting a value of this random variable equal to 0.713 or less is exactly 0.713.

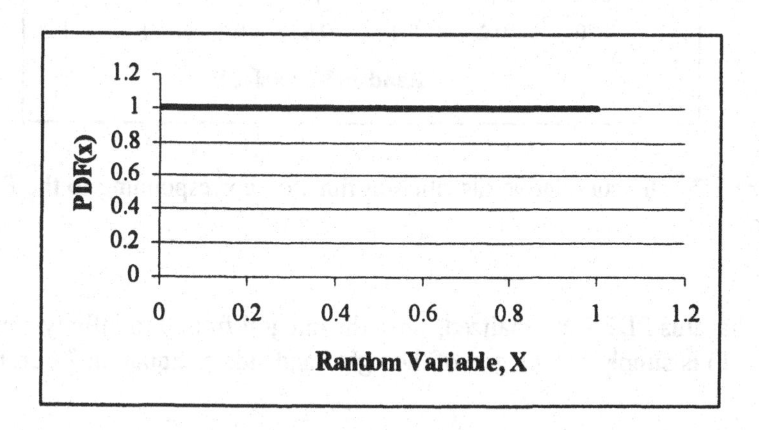

Figure 7.2. A uniform PDF defined on the interval [0,1]. Note that any value of X has an equal probability of being selected in a random draw.

Imagine selecting a value at random from this uniform PDF (e.g. 0.713). Consider next the PDF for one of the parameters for which a random value must be selected for the quadruplet (e.g. the PDF for the source term, ST). We might ask: *Which value of the random variable ST has the same value of the CDF as that selected from the PDF for the uniform distribution*? For example, which value of ST has a cumulative distribution equal to 0.713? If we can identify this value, we can substitute this value of ST into the defining equation (Equation 7.1) for the first realization of P.

It is necessary, therefore, to convert the PDFs for the four parameter values in Equation 7.1 (or Premises 1, 2, 4 and 7 in Figure 7.1) into CDFs. As a simple example, consider the case in which the PDF for ST is exponential:

$$(7.6) \qquad \mathrm{PDF(ST)} = \exp(-\mathrm{ST})$$

where exp is the exponential function.

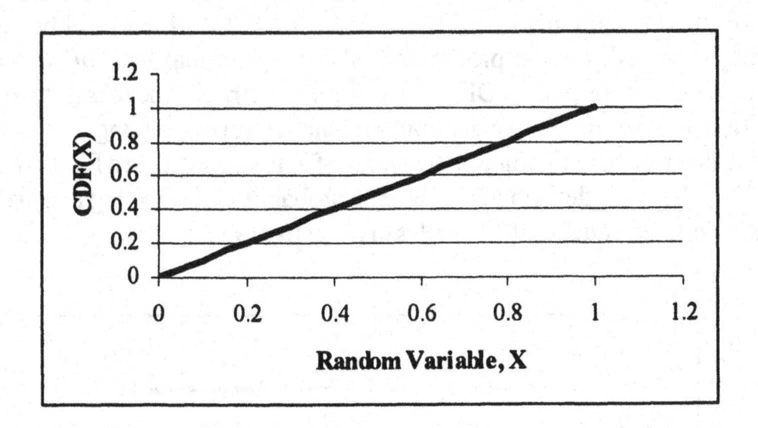

Figure 7.3. The cumulative distribution function corresponding to the PDF in Figure 7.2.

Note that this PDF is normalized, since the integral from 0 to infinity equals 1. CDF(ST) is simply the integral of the right hand side of Equation 7.6 from 0 to ST:

$$(7.7) \qquad \mathrm{CDF(ST)} = \int \exp(-\mathrm{ST})\, d\mathrm{ST} = 1-\exp(-\mathrm{ST})$$

Which value of ST has the same CDF value as the random variable selected from the uniform PDF (remembering that the latter value was 0.713)? It is precisely the value of ST for which CDF(ST) equals 0.713. This yields:

$$(7.8) \qquad \mathrm{CDF(ST)} = 0.713 = 1-\exp(-\mathrm{ST})$$

or

$$(7.9) \qquad \mathrm{ST} = -\mathrm{LN}(1-0.713) = -\mathrm{LN}(0.287) = 1.25 \text{ g/s}$$

where LN is the natural logarithm. In general, if a value from the uniform distribution equal to X is obtained, the corresponding value of ST characterized by an exponential PDF (with the same CDF value as the random number) is:

(7.10) $$ST = -LN(1-X)$$

Note: Equation 7.10 is valid only for an exponential PDF; analogous equations could be defined for other PDFs.

This process may then be repeated for each of the four parameter values in the quadruplet (ST, DC, ER, SF). The result is the following sequence of steps:

1. Select a value at random from the uniform PDF (let this be X_1)

2. Locate the corresponding value of ST from the CDF(ST) by a process equivalent to Equations 7.7 through 7.10 with X replaced by X_1 in Equation 7.10

3. Select a second value at random from the uniform PDF (let this be X_2)

4. Locate the corresponding value of DC from the CDF(DC) by a process equivalent to Equations 7.7 through 7.10 with X replaced by X_2 in Equation 7.10

5. Select a third value at random from the uniform PDF (let this be X_3)

6. Locate the corresponding value of ER from the CDF(ER) by a process equivalent to Equations 7.7 through 7.10 with X replaced by X_3 in Equation 7.10

7. Select a fourth value at random from the uniform PDF (let this be X_4)

8. Locate the corresponding value of SF from the CDF(SF) by a process equivalent to Equations 7.7 through 7.10 with X replaced by X_4 in Equation 7.10

9. Use these four values in Equation 7.1 to obtain the first realization of P

10. Repeat the above process (Steps 1-9) N times to obtain N realizations of the value of P

11. Summarize these N realizations as PDF(P) and/or CDF(P)

The resulting PDF(P) is the probability density function describing intersubject variability of the probability within the exposed population.

In the above process, a new value from the uniform PDF was selected in generating each of the four parameter values required in Equation 7.1. This is because complete independence was assumed between the parameters. If the parameters were fully, positively, correlated (see Chapter 2), a single value from the uniform distribution would have been selected and used to locate all four parameter values:

1. Select a value at random from the uniform PDF (let this be X_1)
2. Locate the corresponding value of ST from the CDF(ST) by a process equivalent to Equations 7.7 through 7.10 with X replaced by X_1 in Equation 7.10
3. Locate the corresponding value of DC from the CDF(DC) by a process equivalent to Equations 7.7 through 7.10 with X replaced by X_1 in Equation 7.10
4. Locate the corresponding value of ER from the CDF(ER) by a process equivalent to Equations 7.7 through 7.10 with X replaced by X_1 in Equation 7.10
5. Locate the corresponding value of SF from the CDF(SF) by a process equivalent to Equations 7.7 through 7.10 with X replaced by X_1 in Equation 7.10
6. Use these four values in Equation 7.1 to obtain the first realization of P
7. Repeat the above processes N times to obtain N realizations of P
8. Summarize these N realizations as PDF(P) and/or CDF(P)

The above process was simple because there is a closed-form solution to the integral of the exponential PDF (Equation 7.8). This is not true in general, since the PDF for a parameter value may be quite complex, particularly if it is obtained empirically rather than being one of the more common parametric forms (exponential, lognormal, probit, etc). In these cases, it will be necessary to integrate the PDF for a parameter numerically (see Chapter 6). The process for a Monte Carlo analysis with a single random variable (here, ST) then is:

Example 7.2. Consider the case in which Equation 7.1 is used to calculate P, and where the PDFs for intersubject variability in ST, DC, ER and SF all are exponential and are independent. Following the procedure outline above, 4 random numbers are generated from the uniform PDF defined on the interval [0,1]. These are 0.2, 0.5, 0.3 and 0.7. For the first realization of P, the following values of ST, DC, ER and SF are obtained:

$ST_1 = -LN(1-0.2) = 0.22$

$DC_1 = -LN(1-0.5) = 0.69$

$ER_1 = -LN(1-0.3) = 0.36$

$SF_1 = -LN(1-0.7) = 1.2$

The corresponding value of P for this first realization is:

$P_1 = 0.22 \bullet 0.69 \bullet 0.36 \bullet 1.2 = 0.066$

Four new random numbers then are generated from the uniform PDF defined on the interval [0,1]. These are 0.4, 0.1, 0.2 and 0.3. For the second realization of P, the following values of ST, DC, ER and SF are obtained:

$ST_2 = -LN(1-0.4) = 0.51$

$DC_2 = -LN(1-0.1) = 0.11$

$ER_2 = -LN(1-0.2) = 0.22$

$SF_2 = -LN(1-0.3) = 0.36$

The corresponding value of P for this second realization is:

$P_2 = 0.51 \bullet 0.11 \bullet 0.22 \bullet 0.36 = 0.0044$

For these two realizations, the CDF is characterized by a value of 50% at 0.01 (since half of the realizations are at this value or less) and 100% at 0.1 (since all of the values are below 0.1). This CDF is not, of course, stable due to the small sample size.

1. Select a value at random from the uniform PDF (let this be X)

2. Locate the corresponding value of ST from the PDF(ST) by numerically integrating PDF(ST) to progressively higher values of ST until the integral equals X; this value of ST then is the random value to be used in the calculation of P

3. Use this value in Equation 7.1 to obtain the first realization of P

4. Repeat the above processes N times to obtain N realizations of P

5. Summarize these N realizations as PDF(P) and/or CDF(P)

To simplify the process, the numerical integration might be performed prior to the Monte Carlo process and summarized as a table with ST in the first column and CDF(ST) in the second column. This removes the need to re-perform the integration for each realization. Step 2 above then is replaced by a look-up process in which the second column is searched downwards until a value equal to X is located, and the corresponding value of ST in the first column is selected. Alternatively, it may be possible to fit the tabular results of the numerical integration with a simple polynomial in ST. Once the value of X is selected in step 1 above, the polynomial is equated to X and solved for ST. Both of these approaches avoid the need to perform the numerical integration anew for each of the N realizations.

What value of N is needed in a Monte Carlo analysis? While there is no general answer, there are guidelines that may be applied. Note that the PDF and/or CDF for the prediction of P will change as N increases, especially when N is initially small. Eventually, the PDF or CDF will approach an asymptotic solution for very large values of N. Once this asymptote has been reached, there is no need to generate further realizations. The choice of N, therefore, depends on how close one wishes to approach the asymptotic solution.

It should also be noted, however, that the value of N required to approach an asymptotic solution depends on the property of the PDF or CDF considered. For points near the median of the PDF for P (i.e. points in the central region of the PDF), the value of N required to stabilize the PDF may be relatively small. By contrast, the value of N required to stabilize the PDF in the tails (i.e. for values of P far from the median) may be much larger since only a small fraction of the realizations produce values this far from the median. The choice of N, therefore, depends on the part of the PDF for which stable solutions are sought.

In risk assessment, one usually is interested in values of P (or any other quantity being calculated) that represent the upper tails of distributions. For example, consider the policy goal noted in Section 7.1. The value of Y in that case was 1%, indicating that one wants a stable answer to the variability CDF(P) at the 99th percentile (i.e. a stable answer to the question: *What value of P corresponds to the 99th percentile of the intersubject variability CDF for P?*). The process for selecting N may be summarized as:

1. Perform the Monte Carlo analysis with some starting value of N (this is almost always 1000 or larger)

2. From this analysis, determine the value of the prediction (e.g. P) corresponding to the desired percentile of the CDF (e.g. 99th percentile); call this P_1

3. Increase N by a factor of 2 and repeat the analysis

4. From the new analysis, determine the value of the prediction corresponding to the desired percentile of the CDF (e.g. 99th percentile); call this P_2

5. Calculate $(P_1-P_2)/P_1$

6. Compare $(P_1-P_2)/P_1$ against a goal (e.g. that the difference not exceed 0.05)

7. If the goal is exceeded, increase N by another factor of two and repeat the above process, now using $(P_2-P_3)/P_2$

8. If the goal is not exceeded, consider the resulting PDF(P) or CDF(P) to be stable

7.4. Incorporating Model Uncertainty

The methods of Section 7.3 may be applied to uncertainty analysis, so long as that analysis is conditional (i.e. does not reflect uncertainty in model form, but only uncertainty in parameter values). The sole differences between variability and uncertainty analyses are that (i) the PDFs for the parameters will now characterize uncertainty in the mean value of the parameter within the population and (ii) the PDF for the predictions will now characterize the uncertainty in the mean value of the probability of effect within the population.

Note that in Figure 7.1, it did not make sense to consider variability in the model form. The reason is that the phenomena being modeled do not possess multiple model forms, but rather are governed by the laws embodied in a single model. There is uncertainty in the model form, but not variability. As a result, variability analysis is always conditional upon a given model. It is common to perform the variability analysis using the best available model, and to then consider competing models in uncertainty analysis.

What is to be done in uncertainty analysis if there is uncertainty in not only the parameter values, but also the model into which these parameters are placed? Several principles of rationality suggest themselves [4]:

- One might select the best available model and perform a conditional uncertainty analysis using this model in a procedure identical to that discussed in Section 7.3, using PDFs for uncertainty rather than variability.

- One might select the model that is most conservative (i.e. the model that is likely to lead to the highest predictions of risk) and again perform a conditional uncertainty analysis. This will ensure that, if the model is incorrect, it will at least lead to overestimates of risk rather than underestimates.

- One might formally consider the uncertainty in models, reflecting this uncertainty in the final PDF or CDF produced for the predictions of risk.

The present section considers the third option above.

It is necessary first to consider the criteria by which a model is to be judged, and how these criteria are to be used in generating a measure of model confidence. In Figure 7.1, there are two models postulated for the exposure-response relationship (a linear model on the left and a quadratic model on the right). What is the relative amount of confidence to be assigned to each of these two models? To address this issue, consider the following criteria for judging the success of a model [5]:

- *Empirical success*, or the ability of a model to predict correctly the results of experiments or other forms of observation. Empirical success involves measures of goodness of fit to data. A relevant measure of the relative empirical success of two models is the relative likelihood. A body of data is obtained and used as the critical empirical test of the models. For each of the models, the best-fitting parameters are obtained and the likelihood of obtaining the data conditional upon the model being correct is calculated. For example, the likelihood might be 0.02 for the linear model and 0.01 for the quadratic model (these values will be used throughout this section as an example). The relative likelihood then is 0.02/(0.02 + 0.01) = 0.67 for the linear model and 0.01/(0.02 + 0.01) = 0.33 for the quadratic model. These two values are then equated with the confidence in each model.

- *Conceptual success*, or the ability of the model to rest on assumptions that are demonstrated to be correct descriptions of the underlying processes, and the ability of the model to explain key phenomena of interest. Conceptual success represents scientific *understanding*, while empirical success represents *prediction* [6]. In other words, empirical success is concerned with getting the right quantitative answer, while conceptual success is concerned with getting the right answer for the right reason. Measures of conceptual success are necessarily subjective and do not involve statistical methodologies.

Empirical and conceptual measures of model success might be combined in one of two ways. One approach is to produce a fully subjective measure based on reflection of both of the above measures of success. For example, if the relative likelihood of Models 1 (the linear model) and 2 (the quadratic model) were 0.67 and 0.33, respectively, and Model 1 had greater conceptual success, the overall measure of confidence might be increased for Model 1 (perhaps from 0.67 to 0.8) and the confidence in Model 2 decreased (to 0.2 from 0.33).

Bayesian methods might also be used to combine the two measures. Consider the state of confidence prior to fitting the model to data. The assumptions underlying model 1 might be considered more scientifically realistic than those underlying model 2. As an example, imagine a subjective measure of confidence (due to conceptual success) of 0.6 for Model 1 and 0.4 for Model 2. This subjective measure will be the *prior* for each model (prior because it is formed prior to fitting the model to the data on which likelihood will be based). The likelihood for Model 1 will be 0.02 and for Model 2 will be 0.01. Using the Bayesian relationship [7]:

$$C_i = P_i \bullet L_i / \Sigma (P_j \bullet L_i) \quad (7.11)$$

where C_i is the confidence in Model i, P_i is the prior for Model i, L_i is the likelihood for Model i (conditional upon the data selected as the empirical test) and the summation in the denominator is over all models j. It should be noted that Equation 7.11 provides only a relative measure of confidence for the different models available, and not an absolute measure of confidence. The absolute measure of confidence probably is lower (by some unknown amount), since the denominator should contain all possible models but usually contains only the existing models. Bayesian measures of confidence are intermediate between purely subjective measures and more statistical measures.

Example 7.3. Using the prior and likelihood for Models 1 and 2 mentioned above, the Bayesian confidence for each model is:

$$C_1 = 0.6 \bullet 0.02 / (0.6 \bullet 0.02 + 0.4 \bullet 0.01) = 0.75$$

$$C_2 = 0.4 \bullet 0.01 / (0.6 \bullet 0.02 + 0.4 \bullet 0.01) = 0.25$$

How may the uncertainty analysis formally reflect these measures of confidence in the models? One approach is to perform the uncertainty analyses conditional upon the first model, and then to repeat the analysis conditional upon the second model. Note that from Figure 7.1, the two analyses would use the same uncertainty PDFs for ST, DC and ER; the analysis for Model 1 would use the linear model and the uncertainty PDF for the slope factor (SF); and the analysis for Model 2 would use the quadratic model and the uncertainty PDF for the quadratic slope factor (SF_Q). For each of these two analyses, the CDF for the prediction of P (the mean probability of effect in the population) is generated. A weighted average of these two CDFs then is developed, with the weighting being the relative measures of confidence in the models.

As an example, consider the case in which the uncertainty in ST, DC, ER and SF produces the CDF(P) shown in Figure 7.4. The confidence in this linear model is 0.75 (as calculated in Example 7.3). The uncertainty in ST, DC, ER and SF_Q produces the CDF(P) shown in Figure 7.5. The confidence in this quadratic model is 0.25 (again, from Example 7.3). To create the composite CDF reflecting uncertainty in P:

1. Select a value on the x-axis (e.g. 0.2)

2. Determine the value of CDF(P) for the first model at this value on the x-axis (this cumulative confidence is 0.63)

3. Determine the value of CDF(P) for the second model at this same value on the x-axis (this cumulative confidence is 0.86)

4. Obtain the weighted average of these values (0.63 • 0.75 + 0.86 • 0.25 = 0.69)

5. Select a new value on the x-axis (e.g. 0.6)

6. Determine the value of CDF(P) for the first model at this new value on the x-axis (this cumulative confidence is 0.95)

7. Determine the value of CDF(P) for the second model at this new value on the x-axis (this cumulative confidence is 0.998)

8. Obtain the weighted average of these values (0.95 • 0.75 + 0.998 • 0.25 = 0.962)

9. Repeat the above process for other values of P on the x-axis, generating the weighted CDF(P)

An example of the weighted CDF using the CDFs in Figure 7.4 and 7.5, and weightings of 0.75 and 0.25, respectively, is shown in Figure 7.6.

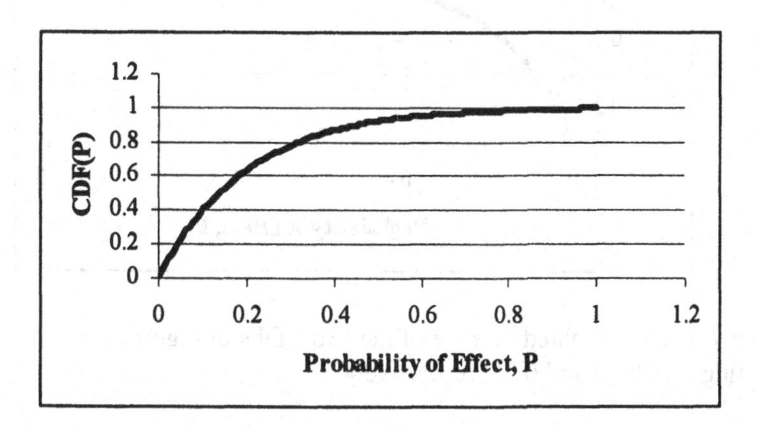

Figure 7.4. A hypothetical CDF reflecting uncertainty in the probability of effect for Figure 7.1, using the linear model. The confidence in this model is 0.75.

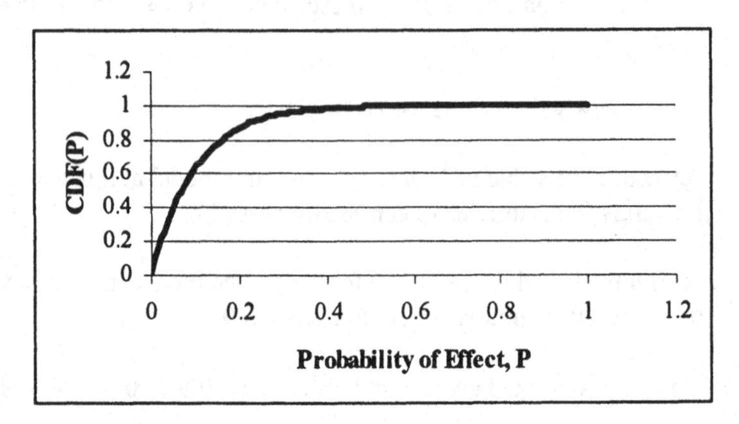

Figure 7.5. A hypothetical CDF reflecting uncertainty in the probability of effect for Figure 7.1, using the quadratic model. The confidence in this model is 0.25.

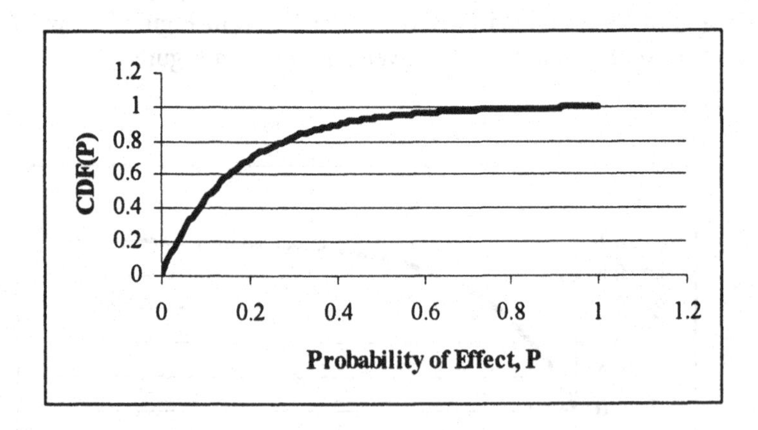

Figure 7.6. The weighted average of the two CDFs in Figures 7.4 and 7.5. The weightings are 0.75 and 0.25, respectively.

An equivalent approach to generating the composite CDF, and one that does not involve generation of the two separate CDFs followed by weighting (but yields the same CDF), is to incorporate the selection of the model directly into the Monte Carlo analysis. The entire procedure is described below for the case of Equation 7.1 and where the uncertainty PDFs for the parameters (i.e. the uncertainty in the mean value for the exposed population) have been generated and converted to uncertainty CDFs. The procedure is very similar to that used in Section 7.3, with the exception of the steps involving confidence in models. In this example, the uncertainty PDFs are assumed to be exponential, but the method is not restricted to such forms:

1. Select a value at random from the uniform PDF (let this be X_1)

2. Locate the corresponding value of ST from the uncertainty CDF(ST) by a process equivalent to Equations 7.7 through 7.10 with X replaced by X_1 in Equation 7.10

3. Select a second value at random from the uniform PDF (let this be X_2)

4. Locate the corresponding value of DC from the uncertainty CDF(DC) by a process equivalent to Equations 7.7 through 7.10 with X replaced by X_2 in Equation 7.10

5. Select a third value at random from the uniform PDF (let this be X_3)

6. Locate the corresponding value of ER from the uncertainty CDF(ER) by a process equivalent to Equations 7.7 through 7.10 with X replaced by X_3 in Equation 7.10

7. Select a fourth value at random from the uniform PDF (let this be X_4)

8. Locate the corresponding value of SF from the uncertainty CDF(SF) by a process equivalent to Equations 7.7 through 7.10 with X replaced by X_4 in Equation 7.10

9. Use these four parameter values (of ST, DC, ER and SF) in Equation 7.1 to obtain the first realization of P under the linear model; call this P_{1L}

10. Select a fifth value at random from the uniform PDF (let this be X_5)

11. Locate the corresponding value of SF_Q from the uncertainty CDF(SF_Q) by a process equivalent to Equations 7.7 through 7.10 with X replaced by X_5 in Equation 7.10

12. Use these four parameter values (of ST, DC, ER and SF_Q) in the equation $P = SF_Q \bullet (ST \bullet DC \bullet ER)^2$ to obtain the first realization of P under the quadratic model; call this P_{1Q}

13. Select a sixth value at random from the uniform PDF (let this be X_6)

14. If X_6 is less than C_1 (i.e. less than 0.75), multiply P_{1L} by 1 and P_{1Q} by 0 and sum these two products; otherwise, multiply P_{1L} by 0 and P_{1Q} by 1 and sum these two products; the sum is the first realization of P

15. Repeat the above processes N times to obtain N realizations of P

16. Summarize these N realizations as PDF(P) and/or CDF(P)

Note that through this process, Model 1 is selected as the correct model in a fraction C_1 of the realizations, and Model 2 is selected as the correct model in a fraction C_2 of the realizations. The procedure can be extended easily to a larger number of models. Also, the resulting CDF(P) will be the same as that obtained under the first method discussed in this section (involving a weighted average of the two conditional CDFs produced under the two models).

7.5. Variability Between Geographic Regions and Subpopulations

The second procedure described in Section 7.4 provides a useful means to combine intersubject variability distributions for different subpopulations. These subpopulations might be distinct because they are in different geographic regions (e.g. different grid blocks) and/or possess different exposure characteristics and susceptibilities/sensitivities.

As in the case of uncertainty analysis, consider a population divided into two subpopulations, with a fraction F_1 (e.g. 0.75) of the people in the first subpopulation and a fraction F_2 (e.g. 0.25) in the second subpopulation. These two subpopulations are characterized by different intersubject variability PDFs for ST, DC, ER and SF (only the linear model is considered here, since it is the model judged in the example of Section 7.4 to have the greatest confidence). To generate the overall variability PDF(P) or CDF(P) for the combined population (including both subpopulations), the following procedure may be used:

1. Select a value at random from the uniform PDF (let this be X_1)

2. Locate the corresponding value of ST from the variability CDF(ST) for the first subpopulation by a process equivalent to Equations 7.7 through 7.10 with X replaced by X_1 in Equation 7.10

3. Select a second value at random from the uniform PDF (let this be X_2)

4. Locate the corresponding value of DC from the variability CDF(DC) for the first subpopulation by a process equivalent to Equations 7.7 through 7.10 with X replaced by X_2 in Equation 7.10

5. Select a third value at random from the uniform PDF (let this be X_3)

6. Locate the corresponding value of ER from the variability CDF(ER) for the first subpopulation by a process equivalent to Equations 7.7 through 7.10 with X replaced by X_3 in Equation 7.10

7. Select a fourth value at random from the uniform PDF (let this be X_4)

8. Locate the corresponding value of SF from the variability CDF(SF) for the first subpopulation by a process equivalent to Equations 7.7 through 7.10 with X replaced by X_4 in Equation 7.10

9. Use these four values (of ST, DC, ER and SF) from Steps 1-8 in Equation 7.1 to obtain the first realization of P for the first subpopulation; call this P_{11}

10. Select a fifth value at random from the uniform PDF (let this be X_5)

11. Locate the corresponding value of ST from the variability CDF(ST) for the second subpopulation by a process equivalent to Equations 7.7 through 7.10 with X replaced by X_5 in Equation 7.10

12. Select a sixth value at random from the uniform PDF (let this be X_6)

13. Locate the corresponding value of DC from the variability CDF(DC) for the second subpopulation by a process equivalent to Equations 7.7 through 7.10 with X replaced by X_6 in Equation 7.10

14. Select a seventh value at random from the uniform PDF (let this be X_7)

15. Locate the corresponding value of ER from the variability CDF(ER) for the second subpopulation by a process equivalent to Equations 7.7 through 7.10 with X replaced by X_7 in Equation 7.10

16. Select an eighth value at random from the uniform PDF (let this be X_8)

17. Locate the corresponding value of SF from the variability CDF(SF) for

the second subpopulation by a process equivalent to Equations 7.7 through 7.10 with X replaced by X_8 in Equation 7.10

18. Use these four values (of ST, DC, ER and SF) from Steps 10-17 in Equation 7.1 to obtain the first realization of P for the second subpopulation; call this P_{12}

19. Select a ninth value at random from the uniform PDF (let this be X_9)

17. If X_9 is less than F_1 (i.e. less than 0.75), multiply P_{11} by 1 and P_{12} by 0 and sum these products; otherwise, multiply P_{11} by 0 and P_{12} by 1 and sum these products; the sum is the first realization of P

18. Repeat the above processes N times to obtain N realizations of P

19. Summarize these N realizations as PDF(P) and/or CDF(P)

The result will be the composite PDF(P) and/or CDF(P) for the entire population including both the first and second subpopulations. Again, the procedure can be extended to any number of subpopulations so long as the fraction of people in each subpopulation and the variability distributions for the different subpopulations can be obtained.

7.6. Nested Variability and Uncertainty Analysis

Consider in this last section the policy goal stated in Section 7.1, and which involves recognition of both variability and uncertainty:

Find a policy (e.g. control option on an industrial facility) that will ensure with at least X% confidence (e.g. 95% confidence) that no more than Y% of the population (e.g. 1%) will have a risk exceeding Z (e.g. 10^{-4}).

Locating such a policy requires performing both an uncertainty and a variability analysis. Using again the logic tree in Figure 7.1, consider the case in which there is a single population of people, with intersubject variability in ST, DC, ER and SF (uncertainty in model form is not considered here, but may be incorporated using methods from Section 7.4). For each of the four parameters, there is a PDF representing the variability distribution, characterized by a mean, M, and variance, V. For example, $PDF_{var}(ST)$ has mean $M_{var,ST}$ and variance $V_{var,ST}$. The subscript var indicates that the distributions refer to intersubject variability.

In addition, there is uncertainty in the values of M_{var} that characterize the variability distributions. This uncertainty (in the mean value of the variability distribution) is characterized by a PDF shown as $PDF_{unc}(ST)$, $PDF_{unc}(DC)$, etc, where the subscript unc indicates that these are uncertainty distributions and not variability distributions. The task is now to perform a variability analysis nested inside an uncertainty analysis. In general, this means that the uncertainty PDFs are used to select at random the distributional characteristics of the variability distributions for the parameters in the model; these variability distributions are used to produce the variability distribution for P, summarized as $PDF_{var}(P)$ and/or $CDF_{var}(P)$; and this process is repeated many times to produce N versions of $PDF_{var}(P)$ and/or $CDF_{var}(P)$. The procedure is as follows:

1. Produce the best estimate of the PDF describing inter-subject variability for each parameter value (ST, DC, ER and SF) that will be used in the calculations shown in Equation 7.1

2. For each of these four PDFs, determine the uncertainty in the mean value of the distribution (we will ignore uncertainty in the variance here, although that also may be incorporated using the same methods). Characterize this uncertainty by $PDF_{unc}(ST)$, $PDF_{unc}(DC)$, $PDF_{unc}(ER)$ and $PDF_{unc}(SF)$

3. Using $PDF_{unc}(ST)$, select randomly one mean to be applied to $PDF_{var}(ST)$; this step (and the steps below) uses the methods discussed in Sections 7.3 and 7.4 based on the uniform PDF for the interval [0,1]

4. Using $PDF_{unc}(DC)$, select randomly one mean to be applied to $PDF_{var}(DC)$

5. Using $PDF_{unc}(ER)$, select randomly one mean to be applied to $PDF_{var}(ER)$

6. Using $PDF_{unc}(SF)$, select randomly one mean to be applied to $PDF_{var}(SF)$

7. Having established the necessary variability PDFs for the parameters, conduct a full variability Monte Carlo analysis using the procedures of Section 7.3. In each realization of this variability analysis, the variability PDFs for the four parameters remain fixed.

8. Produce the variability distribution describing the inter-subject variability of risk. This one PDF(P) or CDF(P) was obtained using the specific variability PDFs generated above, with their specific means.

9. Now repeat the entire process by selecting a new set of means for the four variability PDFs (one for each of the four parameters) using the four uncertainty PDFs. Produce a new variability distribution for risks in the population.

10. Repeat all of the steps N times. The result will be a set (of size N) of variability distributions such as those shown in Figure 7.7.

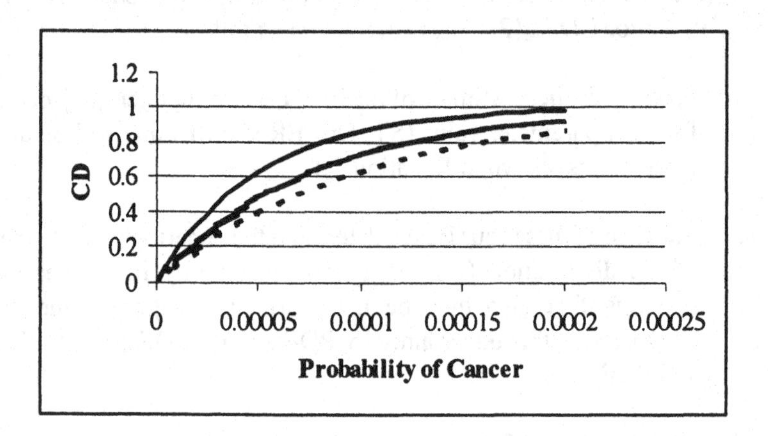

Figure 7.7. A hypothetical example of three realizations of an intersubject variability distribution, based on three different random selections (from uncertainty distributions) of the means for the variability distributions of the underlying parameters.

A nested variability and uncertainty analysis such as the one shown in Figure 7.7 may then be used to address the policy question raised at the beginning of this section. We may ask: *With what confidence can one state that the fraction of people with a probability of cancer below 10^{-4} is less than or equal to Y%?* From Figure 7.7, we see that the three estimates of the fraction of people with a probability of cancer below this limit are 0.6, 0.7 and 0.9. We are, therefore, 33% confident that the fraction is at or below 0.6; 67% confident that it is at or below 0.7; and 100% confident that it is at or below 0.9. Clearly, we are not very confident that the fraction of people whose risk is below 10^{-4} does not exceed 1% (the fraction stated in the policy goal), and so this policy would not be accepted. Of course, with a larger number of runs through the same process, we would produce much finer resolution on this information.

On a final note, there is a tendency at times to conflate uncertainty and variability. The argument usually is that variability in parameters values within a population is simply another source of uncertainty about the correct parameter value to apply to a specific individual. Such arguments are based on the idea that risk assessment is concerned with selecting an individual at random from the population and attempting to estimate the risk to that individual. In this case, it certainly is true that, in the absence of information allowing the selection of an individual-specific parameter value (such as a slope factor specific to an individual), the presence of intersubject variability introduces uncertainty into selection of a parameter value to be applied to this randomly selected individual.

This argument is not valid, however, for two main reasons. First, risk assessment is not simply concerned with the random selection of individuals and the assignment of risks to these. It is, instead, also concerned with characterizing the actual variation of risk across a population. It is important to bear in mind that this variation in risk is an objective property of the population, and not due simply to uncertainty. The decision-maker needs information on this variation, so the variation can be compared against such secondary goals as risk equity (an issue of environmental justice).

Second, uncertainty can be reduced through further research, while intersubject variability cannot. The actual variation in risk across individuals in a population can be characterized truthfully or poorly, but it cannot be changed by further research. By contrast, uncertainty can be reduced through research. It is essential in risk assessment, therefore, to clearly demarcate variability and uncertainty, and to perform nested uncertainty/variability analyses. To do otherwise is likely to confuse the decision-maker, since it will fail to differentiate between two distinctly different aspects of decisions.

References

1. Hoffman, F. and Gardner, R., “Evaluation of Uncertainties in Radiological Assessment Models”, in *Radiological Assessment: A Textbook on Environmental Dose Analysis*, ed. by Till, J. and Meyer, H., U.S. Nuclear Regulatory Commission, 1983.
2. for examples of such analyses, see C. Cothern and J. Smith, *Environmental Radon*, Plenum Press, New York, 1987.
3. J. Gentle, *Random Number Generation and Monte Carlo Methods*, Springer Verlag, New York, 1998.
4. D. Crawford-Brown, *Risk-Based Environmental Decisions: Methods and Culture*, Kluwer Aademic Publishers, Boston, 1999.
5. D. Crawford-Brown and J. Arnold, "Theory Testing, Evidential Reason and the Roleof Data in the Formation of Rational Confidence Concerning Risk", *Comparative Environmental Risk Assessment*, ed. by C. Cothern, Lewis Publishers, 1992.
6. S. Toulmin, *Foresight and Understanding*, Hutchinson, London, 1961.
7. J. Bernardo and A. Smith, *Bayesian Theory*, John Wiley and Sons, New York, 2000.

INDEX

Additivity of fields, 19
Aggregate exposure, 143-144
Analytic methods, 176-177
Augmented matrix, 134
Average daily rate of intake, 153

Bayes' theorem, 33, 189
Bayesian, 33
Bernoulli's method, 74-93
Binomial distribution, 43-44
Biologically-based models, 150
Buoyancy, 15

Carriage, 15-16
Catenary systems, 73
Central tendency, 36
Change of scale property, 100
Class of solutions, 68
Closed system, 64
Co-factor, 136-138
Columns of matrices, 129
Compartment, 62, 168
Conceptual success, 189
Confidence, 31
Connectors, 168
Conservation of energy, 66-67
Conservation of mass, 66-67
Continuous variable, 35
Correlation, 46-48
Coupled systems, 95
Crosswind walk, 17
Cumulative distribution function, 39-40, 179, 181
Curl, 13-14

Darcy's law, 8
Descriptive statistics, 29
Determinant, 136-140
Diagonal elements, 132
Differential equation, 68
Diffusion, 16
Discrete variable, 35
Dispersion coefficient, 18
Divergence, 9-11
Downwind walk, 17

Effective stack height, 18
Empirical success, 188-189
Environmental tobacco smoke, 3-4
Equilibrium, 80
Equivalent rectangles, 154
Error propagation, 56-59
Euler's method, 157-163
Exponential washout, 76-78
Exposure-response, 147

Feedback 95-96
Field, 2
Field equation, 3
Field evolution, 3
Field property, 2
Field quantity, 2
First-order process, 70
Frequency, 30

Gaussian back-elimination 126-128
Gaussian plume model, 17-25
Gauss-Jordan method, 134-136
Geometric standard deviation, 42, 177
Gradient, 6-9
Gridded search, 53

Half-life, 78
Heaviside expansion theorem, 102

Histogram, 37
Homeostasis, 96
Homogeneous field, 2

Identity matrix, 132
Inferential statistics, 29
Inhalation rate, 61-63
Inhomogeneous field, 2
Initial condition, 69
Intersubject variability, 175
Inverse distribution function, 40-41
Inverse Laplace transform, 101-102
Inverse matrix, 134
Inverse transformation, 97
Isopleth, 9

Laplace transform, 95-123
Least squares, 49-52
Likelihood, 31, 34
Linear dependence, 128
Linearity property, 100
Logic tree, 180
Lognormal distribution, 42, 177
Long-term frequency, 32-33

Margin of safety, 28
Matrix, 126, 128-134
Matrix addition, 131
Matrix analysis, 126
Matrix associativity, 132
Matrix commutivity, 131
Matrix identity, 130
Matrix multiplication, 132
Matrix symmetry, 131
Matrix transitivity, 131
Maximum likelihood, 49, 52-53
Mean, 36
Median, 42, 177
Method of descent, 54
Method of substitution, 127
Model uncertainty, 187-194
Monte Carlo methods, 175-199

Negative correlation, 46
Nested analyses, 196-199
Normal distribution, 36, 41
Numerical integration, 152-157
Numerical methods, 147-173
Numerical solution, 147

Off-diagonal elements, 132
Ontological probability, 32
Open system, 64-65, 111

Partitioning, 102, 110
Pearson correlation coefficient, 48
Perturbation, 96, 125
Pharmacokinetics, 98, 149
Pipes, 168
Poisson distribution, 44-45
Positive correlation, 46
Post-multiplication, 130
Precautionary principle, 28
Pre-multiplication, 130
Prior, 189
Probability density function, 35, 37-39
Probability, 29-30
Property, 1

Quantity, 1

Random search, 54
Realization, 179
Regulators, 168
Response, 125
Rotation of fields, 22-23
Rows of matrices, 129
Runge-Kutta methods, 163-167
Run-off, 145-146

Scalar field, 5
Sedimentation, 16
Sensitivity, 125
Shifting property, 100
Simulation, 179
Spectra, 125
Spectral unfolding, 125
Square matrix, 132
State of the environment, 4
State vector, 63

State, 1
State-vector model, 150
State-vector, 150-151
Steady state, 80
STELLA, 167-173
Stochasticity, 28, 175-176
Subcompartment, 84-87
Superposition, 19
System equilibrium, 140-143
System state, 62-63

Tensor field, 5
Threshold model, 150
Time-steps, 158
Transfer rate constant, 70
Transfer, 65
Transformation rate constant, 70
Transformation, 65
Transforms, 97
Translation of fields, 19-20
Transport, 65
Transpose, 133-134
Trapezoidal method, 154-155

Uncertainty PDF, 175
Uncertainty, 28
Uncoupled systems, 73
Unique solutions, 68
Unit, 1

Valves, 168
Variability, 27, 29
Variability PDF, 175
Variance, 36, 56-59
Vector field, 5
Velocity field, 9-11

Watershed, 145
Wind rose, 16, 24

Zero matrix, 132
Zeroth-order process, 69-70
z-score, 47